T0361242

HOME OWNERSHIP IN CRISIS?

Home Ownership in Crisis?
The British experience of negative equity

RAY FORREST
PATRICIA KENNETT
PHILIP LEATHER
School for Policy Studies
University of Bristol

Routledge
Taylor & Francis Group

LONDON AND NEW YORK

Contents

List of figures and tables

Acknowledgements

This book draws on two research projects funded by the Joseph Rowntree Foundation. Further work on negative equity and attitudes to home ownership was carried out for the Department of Employment, Transport and the Regions. The interpretation of the data and views expressed are entirely those of the authors. We should also like to thank everyone in Luton and Bristol who participated in the study.

The entire typescript was prepared for publication by Angela Torrington.

1 Introduction

Writing this book in 1998 it is easy to forget the gloomy and dramatic predictions about the future of the home ownership market in Britain throughout most of the decade. As the recession rolled on the home owner's world seemed to have changed forever. Estate agents' offices sat empty and abandoned in the high streets, conveyancing lawyers lost their jobs and property professionals appeared among the repossession casualties. In public houses and at dinner parties throughout the land the talk was not of price rises but of price falls. It was like one of these social disruptions so beloved by anthropologists when the central tenet of a cult is comprehensively undermined. A belief system, policy framework and institutional structure predicated upon real house price inflation was suddenly knocked sideways. If house prices could fall substantially what could we believe in anymore? Perhaps the world has changed and the current steady if uneven recovery in house prices is merely the precursor to another round of price falls. At the moment, however, it seems more like business as usual.

Negative equity emerged as an important policy issue in the late 1980s in Britain. It also entered popular discourse through television, newspapers and other media. Suddenly, home ownership seemed to lose its untarnished image. There had always been casualties but they were an unpublicised minority of the unlucky or feckless who had over-committed themselves or who could not manage their finances. For the majority, the move from the rental tenures or the parental home to house purchase represented a smooth transition to a varied but generally unproblematic trajectory in the owner occupied sector. Fuelled by a combination of real income growth, financial deregulation, council house sales and a policy environment which had a strong tenurial bias towards house purchase, home ownership grew rapidly throughout the 1980s. By the end of that decade over two thirds of the UK population were in the sector and anticipating a continued real appreciation in property values. When both real and nominal house prices went into rapid decline it seemed that one of the few remaining certainties in life had gone. House prices went up. They didn't go down - at least not in nominal terms and not for long.

The sheer size of the home ownership sector by the end of the 1980s meant that inevitably more people than ever before were exposed to a riskier and more volatile housing market. And unusually, these people were concentrated in the core region of the British space economy, the south east of England where levels of home ownership were particularly high. A concentration of debt encumbered home owners in that part of Britain represented not only personal difficulties for those affected but had important implications for the macro economy. This was the heart of mobility and consumerism. Just as the soaring south eastern housing market in the late 1980s had contributed to a general consumer boom and facilitated the growth of small businesses and self employment, a deep and unprecedented recession in the residential property market was likely to delay any recovery in the broader economy. The capital appreciation in the residential property market which had been the source of secure borrowing for individuals and small businesses had been transformed into a substantial debt overhang.

This book explores this particular period in the history of the British housing market with a special emphasis on the experiences of the households affected by 'negative equity' – where outstanding mortgage debt exceeds the current value of the dwelling. A number of analysts and institutions produced regular estimates and updates of the number of households affected by negative equity in Britain. These are reported and commented on in later chapters. Our research, however, was predominantly at the micro level, endeavouring to gain an understanding of the behavioural and attitudinal impacts of negative equity on individuals and households. Special social surveys and in-depth interviews were combined in three separate research projects carried out between 1992 and 1997. All these studies were concerned with both the way in which individuals and families were negotiating and adapting to an alien economic environment and the macro consequences of these behavioural changes for housing market dynamics and the broader macro economy.

In 1993 the authors were involved in a wide ranging research programme on Housing and the Macro Economy funded by the Joseph Rowntree Foundation. The programme had a number of elements including a social survey of 1,200 home owners in Glasgow, Bristol and Luton. These were three areas which had experienced very different housing market conditions in the 1980s and early 1990s. Glasgow represented a relatively stable housing market in price terms, Bristol was moderately volatile, and Luton was at the more extreme end of the continuum with rapid house price inflation in the late 1980s followed by a steep descent. From the work of Dorling et al. (1992) and from what was known generally about regional house price trends, it was clear that negative equity was a mainly southern phenomenon. Luton, moreover, had been identified as one of the areas with a high concentration of households with negative equity. This survey enabled some broad comparisons to be made between those with negative equity and those without. The survey was also used to generate a sub sample of households in Bristol and Luton for follow up in-depth interviews (see Forrest, Kennett and Leather, 1994 for a full report of this research).

This research aimed to move the debate beyond aggregate estimates and gain a more detailed understanding of what negative equity actually meant for individual households. By exposing a range of circumstances in which negative equity arose, the intention was also to inform debates about the kinds of policy initiatives which could be appropriate.

For example, it was not known whether and in what ways housing market behaviour and attitudes towards home ownership were being affected by negative equity. Indeed, it was assumed that both lenders and borrowers were adjusting to a new experience of home ownership. For households, the question was what form did these adjustments take? And to what extent could they be seen as a longer-term shift in attitudes?

In discussions of negative equity it was also important to keep a sense of proportion and not to exaggerate the impact. A crude count of the likely numbers of households in situations of negative equity concealed a wide range of household circumstances. There was likely, for example, to be a major difference between a household with a high income and a relatively small amount of unsecured housing debt and a household with an insecure and low income and even a modest amount of negative equity. Indeed, it was not possible to predict the scale and severity of the problem from the size of the unsecured debt without knowing about the economic and other circumstances of a household. Households with negative equity were likely to be in a wide range of situations facing different problems and with differential access to the financial and other resources which could provide solutions. Global figures on negative equity could indicate a scale of problem which did not emerge at the level of the household. The in-depth interviews in particular aimed to uncover a wide range of housing and family circumstances and various degrees of negative equity in order to explore the interaction between various factors.

This research was taken further in a follow up project, again funded by the Joseph Rowntree Foundation. The earlier work had indicated a particular concentration of negative equity and associated difficulties on large new build estates of the mid to late 1980s. The boom and affordability pressures in this period had contributed towards the creation of substantial new building, particularly on greenfield sites on urban peripheries. These areas inevitably contained large concentrations of young households and families which had moved into home ownership or traded up at precisely the time when prices peaked and then went into steep decline. It also appeared from the previous work that price falls and difficulties with resale were particularly severe in such areas. Again this research focused on Luton and Bristol but only on two areas of new building - Bradley Stoke adjacent to Bristol on the M4/M5 interchange, and Barton Hills in the north east of Luton. The study involved two main elements. First a social survey of 600 households chosen at random from the postal address files (PAF) in these two locations was carried out by MORI International on our behalf. Second, unstructured follow up interviews were carried out by the present authors with a sub sample of households to pursue in-depth issues that emerged from the social survey. This qualitative work focused on households in both Bristol and Luton which had experienced particular 'events' or were in particular circumstances. This included problems of job mobility, marital breakdown, delayed family formation and job loss, combined with varying levels of positive or negative equity. A total of 37 in-depth interviews were completed (see Forrest, Kennett and Leather, 1997 for a full account of the research method and findings).

The final piece of research in the programme of work on negative equity was commissioned by the Department of Environment, Transport and the Regions (formerly the Department of the Environment) in late 1996. This aimed for a more representative view of the impact of negative equity on household behaviour through the selection of a subsample of households from the English House Condition Survey (EHCS). From a policy perspective a key issue was the impact of negative equity on mobility, spending and savings habits and on attitudes towards dwelling improvement and repair. The EHCS is the major source of data on the physical condition of the housing stock in England but the particular advantage of using the EHCS for a study of negative equity is that unlike most other data sources it includes a professional valuation of properties as opposed to self assessment by home owners as well as a detailed social interview survey with occupants, providing details of outstanding secured loan debt. A total of 284 addresses were identified where households had been in negative or low equity situations in the 1991 survey and a structured questionnaire with a high proportion of open-ended questions was designed. Fieldwork was carried out by MORI International in early 1997 - 157 full interviews were achieved. These data were analysed in an essentially qualitative manner. The survey did not aim for statistical representativeness given the relatively small numbers involved and was originally designed as a precursor to a potentially larger scale study of the impact of negative equity. It was, however, a random sample of households which were identified as having low/negative equity drawn from a nationally representative sample and captured a much wider range of household types and dwellings than previous research based on case studies (Forrest, Kennett and Leather, 1998).

This study was also useful as an exploration of home owners' attitudes and behaviour more generally at a point when the housing market was beginning to recover and as negative equity was diminishing as a problem. The changed economic climate was recognised at the outset and the research became more concerned with the degree to which residual problems would remain if the housing market continued its recovery and how far any changes in attitudes and behaviour in relation to home ownership were likely to be temporary or more deep seated. At the time of the survey uncertainties remained about the scale and pace of the recovery and whether house price inflation would be more uneven than in the past with more evident 'hot spots'. Moreover, some households in particular kinds of dwellings, in particular locations and with particular financial or personal difficulties could remain with a legacy of debt and associated problems. Some of these uncertainties remain. A simple numerical calculation which shows falling numbers of households with negative equity may conceal an array of difficulties which will continue to affect the housing market trajectories of a minority of households. In other words, the experience of negative equity with the stresses and strains which it imposed on some individuals and families does not simply disappear overnight.

This book then draws on evidence from primary research over four years and from other related studies and secondary sources to provide a comprehensive account of the impact of negative equity on the British home ownership market. Given the centrality of home ownership in the national psyche and the ideological baggage which it draws in its wake this is an important period to record. Moreover, the wealth accumulating qualities of home ownership have been an important element of debates about broader social divisions in British society, intergenerational linkages and forms of future welfare provision (see for example, Hamnett et al., 1991; Forrest and Murie, 1995). Falling property values therefore have theoretical and policy ramifications beyond the narrow confines of housing market dynamics and housing policy.

The next chapter provides a brief housing market context within which to situate more specific consideration of negative equity. This is followed by a more focused review of the relevant literature and other studies of negative equity itself, both nationally and internationally. Chapter 4 profiles the findings from secondary analysis of the EHCS to compare households with negative, low and moderate equity in England in 1991. Chapters 5 and 6 present findings from the in-depth interviews carried out in the Bristol and Luton studies in 1994 and 1996. Here we look particularly at different routes into negative equity and the coping strategies adopted by households.[1] Chapter 7 looks more generally at the impact of negative equity on housing market behaviour and attitudes towards home ownership with particular reference to the most recent work carried out for the DETR. The book concludes with an assessment of the British home ownership market and situates the discussion of negative equity and its aftermath in contemporary policy debates about home ownership.

Note

1   A version of Chapter 6 was published in the *Journal of Social Policy*, Forrest, R. and Kennett, P. (1996) 'Housing Careers, Coping Strategies and Households with Negative Equity', pp.369-394.

2 Changing fortunes

For most of the postwar period commentaries on housing problems were confined to the private and public rental sectors. Indeed, until the 1970s both home ownership and council housing were regarded generally as relatively unproblematic tenure forms providing different routes from a declining private rented sector. Low quality housing, problems of indebtedness, insecurity, poverty, overcrowding and other social problems were most strongly associated with private landlordism. As this sector declined new problems arose in the public rental sector. There were increasing references to ghetto estates, residualisation, design and structural faults in system built dwellings, mismanagement, paternalism and excessive bureaucracy. As the poor were increasingly accommodated in an expanding council sector, so inevitably, the tenure began to lose its image as housing for the privileged labour aristocracy, the skilled manual workers of post war Britain. The introduction of the Right to Buy in 1980 combined with an expanding poor and a severely curtailed programme of new building for council renting accelerated this process. More affluent tenants in the most popular council dwellings exited the public rental sector adding a significant new layer to home ownership in Britain (see Forrest and Murie, 1990; Forrest, Gordon and Murie, 1996).

Throughout this period most policy attention was focused on the consequences of the Right to Buy for the tenants and dwellings which remained in the council sector or on other policies designed to facilitate entry to home ownership (see for example, Booth and Crook, 1986). The whole thrust of the Thatcher governments housing policies was to promote home ownership as vigorously as possible. Entry to home ownership was the essential step to obtain membership of an expanding middle class for whom housing equity was pivotal in a broader lifestyle of credit based and

7

housing equity fuelled consumption. Home ownership was a solution not a potential problem. Rates of capital appreciation and the more general experience of residential property ownership were inevitably highly differentiated reflecting broader processes of stratification in society but the experience was generally positive. There were rich home owners in substantial detached houses and the aspirant working class in the Victorian terraces. But there was a pervasive experience of freedom, choice and asset appreciation.

There were, of course, theoretical challenges to this benign picture of a nation of home owners. Some of these critiques were rooted in Marxist theory which situated the growth of suburban home ownership for the middle masses within broader accounts of the changing dynamics of capital accumulation in postwar capitalism, particularly in relation to crises of underconsumption (see Harvey, 1978). From this perspective the growth of home ownership was driven not by rising expectations and changing preferences of consumers but primarily by the relentless search for new areas for profitable exploitation in an increasingly commodified world. Home ownership may have been shaped by grass roots working class movements seeking freedom from private landlords but it was now big business for major financial institutions. Other theoretical excursions linked home ownership to debates about political practice, patriarchy and privatism. These aspects of the debate around the growth of home ownership are engagingly if rather evangelistically reviewed by Saunders (1990) and there is little point in exploring these issues here. Suffice to say that the more arcane and abstract arguments around the social significance of home ownership have proved to be either elusive or illusory or have focused down ultimately on the more empirically observable features of residential property ownership-namely, asset appreciation and house price movements.

Substantial research effort has gone into the measurement of house price movements and differential asset appreciation across different section of the population in a range of national contexts (Forrest and Murie, 1995). An early contribution to this debate, and one of the more theoretically nuanced, was Edel et al. (1984) on the suburbanisation of Boston. They questioned the degree to which home ownership promoted social mobility and argued that the historical experience for many working class Boston suburbanites was of a constant struggle against asset *depreciation* in the context of declining neighbourhoods. For Edel et al. home ownership for some was not so much providing 'a foot on the ladder' as a continual experience of running 'up a down escalator'. Saunders, who dismisses much of the work of Edel et al. and others on differential rates of gain among home owners as "left wing sociological mythology" (p.170) spends much of his analysis on the issue of absolute and relative rates of gain. Much of this debate (and see Forrest et al., 1990 for an alternative view from Saunders) hinged on the degree to which this differentiated experience of house price appreciation united or divided home owning populations. What was generally accepted, however, was that for the majority the lived experience of home ownership was of an appreciating asset and a diminishing debt. Whether that appreciation was real or nominal was a preoccupation of economists rather than an issue which exercised the average household. And there were other commentators who took a more sceptical line and suggested that whatever the evidence on rates of gain in home ownership it was the wrong question to be asking. The point about being a home owner was of *not being a tenant*. Choko (1995), for example, engages with the view that home owners who lose money through buying a home in the wrong location or who could have invested their savings more wisely in other ways are somehow behaving irrationally. "Maybe people do not know whether it is a good investment or not ... Perhaps, since they do not have much choice, most purchasers simply gamble". But Choko goes on to quote Verret (1979) in what he refers to as a more contextualised answer. "To be an owner, or rather, to become an owner ... is above all not to be a tenant anymore. The working class has all the reasons in the world not to want to remain as tenants" (Verret quoted in Choko, 1995, p.147). This observation may be part of the explanation for the resilience of home ownership as the preferred tenure even among households who have had particularly negative experiences. This issue will be returned to in later section when we discuss attitudinal changes among households have had experienced negative equity.

The point of this brief discussion of some of the academic literature on home ownership is to highlight two issues. First, as home ownership has grown in Britain and in many other countries, analysts have become increasingly preoccupied with the movement of house prices and the asset appreciation of residential dwellings compared with other forms of investment. This interest has been promoted by policy concerns with residential mobility, the macro-economic consequences of dwelling equity release and intergenerational transfers and by more theoretical concerns with the sociological significance of potentially new social divisions associated with such forms of wealth. Second, until relatively recently, if there were casualties of the tenure - households which experienced asset depreciation or repossession - they were generally argued to be among the poorer sections of the working class or ethnic minorities typically in depressed regions or declining neighbourhoods. They were *marginal* home owners in terms of income, employment, property or location. Karn et al. (1985), for example, focused on the problems of inner city home ownership in Birmingham and Liverpool. In a piece by Karn, Doling and Stafford (1986) in a collection of essays on The Housing Crisis an account of 'growing crisis and contradiction in home ownership' concludes with reference to the problems which begin to arise when "ownership also covers low income buyers, particularly of inner city terraces" (p.149). And Doling and Stafford (1989) focus on problems of arrears, unemployment and house condition among lower income home owners in Coventry. In other words, problems within the home ownership sector were, until the more generalised recession of the late 1980s, problems on the margins in terms of households and properties. In Britain they were also problems associated with inner cities or declining industrial regions - 'north' rather than 'south'. Commentators searching for emerging problems within the home ownership sector were either forced to make somewhat heroic generalisations about social divisions in the tenure from the specific examples of the experience of minorities or to focus on the more macro level issues of the impact of housing equity on the macro economy. Whether or not the more negative experiences of the late 1980s can be regarded as a fundamental shift in attitudes towards home ownership or in any sense a generalised 'crisis' remains to be explored in later chapters. What is indisputable is that they were unanticipated and affected households and parts of the market previously thought immune from serious asset depreciation. For example, the publication of Saunders' (1990) account of the onward march of home ownership was unfortunate to almost coincide with an unprecedented collapse of property values. And in the same year Lowe (1990) observed that "The rate of capital gains made in housing is closely allied to the point in time of entrance and length of time in the market, so it can be supposed (assuming that there is no catastrophic collapse of the housing market-which seems unlikely) that the spread of owner-occupation of recent years enables relatively new entrants to gear-up in later years" (p.88). Of course, it depends what one means by 'catastrophic' and Saunders and Lowe were by no means alone in focusing primarily on the consequences of widespread and accelerating equity gain as opposed to the problems which could be generated by a rapid shift in the opposite direction. This is hardly surprising given the house price movements of the previous 25 years when average annual real increases had been around 2 per cent. In 1990 most of us were still in the world of assured if uneven house price inflation, high gearing and gazumping - a view of home ownership rooted at least as far back as the boom of the early 1970s when Pawley (1978) remarks that "talk of house purchase as a good investment ceased to be figurative or long term and became instead a matter of immediate importance" (p.132). Falls in nominal house prices were experienced in other countries and debt and repossessions happened to other people. Whatever the longer term significance of the events explored in this book they were a major shock to the housing system and were for some individuals and families unambiguously catastrophic.

A variety of factors contributed to the housing market conditions of the late 1980s in Britain including financial deregulation which improved households' ability to borrow funds for house purchase, changing macroeconomic conditions, government policy and demographic change. These have been fully explored elsewhere (Boleat, 1994, Forrest and Murie 1994, Maclennan et al., 1994) and the remainder of this chapter focuses on some key indicators of change in the residential property market.

The essential background to the changing context for home ownership which emerged in the early 1990s was that it followed a period of unprecedented growth fuelled by house price inflation and associated new building and council house sales. In the previous decade there had been a substantial influx of dwellings and households into the owner occupied sector. In the period 1971 to 1981 home ownership in Britain grew by 8 percentage points. In the next decade the tenure grew by 12 percentage points to encompass more than two thirds of all households. This accelerated growth was most marked in those regions which had previously lagged behind. For example, between 1981 and 1993 the home ownership rate in the North of England grew by 13 percentage points. And Scotland, which in 1980 had two thirds of households in the rental sector, experienced a dramatic tenure transformation over the same period with a 19 percentage point increase in home ownership. By the early 1990s home ownership was the majority tenure in every region in Britain (see Forrest et al., 1995 for a more detailed discussion). There was, however, still a strong north-south dimension to tenure structures with the highest levels of home ownership of 70 per cent or more in the South East (excluding London), the South West and the East Midlands. This is of some importance as house price falls and subsequent negative equity were most severe in precisely those parts of the country with the highest proportion of home owning households.

There is nothing novel about booms and slumps in the housing market. Munro and Tu (1996) observed that "Over the last 25 years there have been three major cycles in real house prices: first, from 1969-1977, peaking in 1973; second from 1977-1982, peaking in 1980 and third from 1982 to the present, with national real house prices still falling from the peak of 1989" (p.1). Moreover, there have been greater falls in real house prices prior to the late 1980s recession. Real house prices fell by 31.4 per cent between 1973 to 1977 compared to 23.8 per cent between 1988 to 1992 (Kennedy and Andersen 1994). However, during the 1970s high inflation rates

11

masked these falls and protected nominal house prices, what Bootle (1996) has referred to as 'the power of inflation to create illusions about real values' (p.68). More recently rates of inflation have been much lower and the significant real house price falls have been accompanied by lower nominal prices (Thomas, 1994; Kempson and Ford, 1995). For Thomas (1996) it is these 'nominal house price declines and resultant incidence of negative equity that marks out the recent housing slump as a uniquely painful episode in the post-war UK housing market' (p.1).

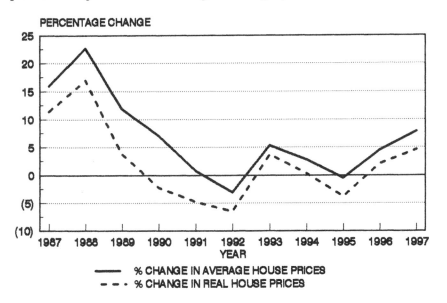

Figure 2.1   **Annual change in average house prices and change in real house prices, UK 1987-1997**

Source: CML Market Briefing.

Figure 2.1 shows the undulations in average house prices and real house prices over the period 1987 to 1997. Changes in real house prices dropped into minus figures in mid 1989 and only began to recover in mid 1992. There then followed a short period of recovery in early 1993 before prices dipped again until mid 1995. The overall severity of the recession and the significant regional variations were well summarised by Boleat (1994). "By far the chief distinguishing feature of the [latest] cycle has been the absolute downturn in house prices, which is without recent precedent. Nominal house prices have fallen by about 30 per cent on average in East Anglia, Greater London, and the South East". He continues "Unlike previous

12

recessions, the current downturn has disproportionately affected England's South and East - areas of the country noted for their traditionally low rate of unemployment, a belief that jobs were always available, and an economy that was always thought to be booming" (pp.258-261). The shock to the system which occurred in the late 1980s was therefore not solely in terms of the depth and severity of the fall in nominal house prices but in terms of its geography. The recession in the early 1980s in Britain was disproportionately northern and focused on the manufacturing sector. The shakeout which began in 1989 was essentially southern and service sector based. It was those regions and the services and financial sector concentrated there which had expanded rapidly in the mid to late 1980s. It was in that region where household borrowing was most highly geared and where self employment had expanded most rapidly. This southern dimension is clearly evident in the pattern of negative equity which emerged.

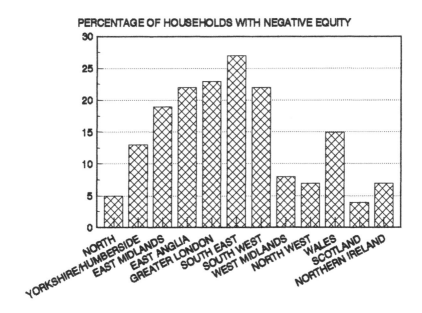

Figure 2.2 The extent of negative equity by region households with negative equity as a proportion of all borrowers – UK 1993

Sources: Department of Employment, Transport and the Regions; Inland Revenue Statistics (1995) Table 5.3, p.66.

As figure 2.2 shows, in 1993 over a quarter of borrowers in the South East had negative equity and over a fifth in London, the South West and East Anglia. By contrast in Scotland and the North of England less than one in 20 borrowers were affected (and see next chapter).

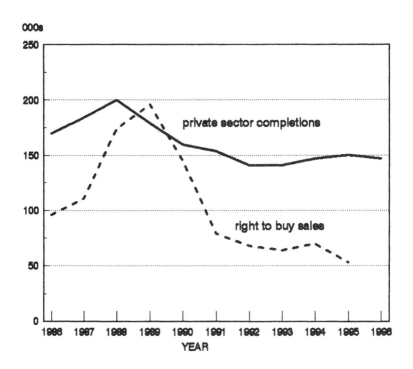

Figure 2.3 Private sector completions and right to buy sales, Great Britain 1986-1996

Source: S Wilcox Housing Finance Review 97/98, Joseph Rowntree Foundation.

The changing state of the market is also revealed by the pattern of building, Right to Buy sales and sales transactions. At the height of the house price boom in 1988 new build completions in the private sector peaked at 207,000. There then followed a continuing and rapid decline. By 1993 private sector completions had fallen by around 30 per cent to 147,000. Similarly, Right to Buy sales which had accounted for almost half of the growth of the tenure in England between 1981 and 1991, fell back dramatically. The late 1980s boom had produced a new surge in sales of

14

public sector dwellings to sitting tenants (there had been an initial high period of selling in England in the early 1980s when the Right to Buy was introduced). In Britain as a whole the combined total of New Town, local authority and housing association sales was 196,000 in 1989. The nature of the Right to Buy makes it highly sensitive to changing conditions. Potential applicants can choose to exercise their Right to Buy or withdraw an application at very short notice. This is shown by the rapid decline to 135,000 RTB completions in 1990 (see figure 2.3). In 1993 there were only 64,000 RTB sales - a third of the 1989 total (Wilcox, 1997). Residential property transactions as a whole in England and Wales were running at almost 2 million in 1988 representing 15 per cent of the stock of owner occupied dwellings and an aggregate value of £120 billion. By 1993, residential transactions had slumped to 1.1 million - just over 7 per cent of all owner occupied properties in England and Wales and with an aggregate value of £67 billion (and see Boleat, 1994).

The late 1980s and early 1990s also saw an unprecedented increase in mortgage arrears and property repossessions. Initially the problems arose mainly with the coincidence of high borrowing, rising job insecurity and high and rising real interest rates. Mortgage interest rates rose from 9.5 per cent in May 1988, the lowest level for ten years, to a peak of 15.4 per cent in 1990. The scale of the increase had a dramatic effect on the housing market and household budgets. The number of properties taken into possession rose from just under 16,000 in 1989 to 44,000 a year later. As interest rates fell unemployment increased and mortgage arrears and possessions continued to rise (see figure 2.4).

In 1992 the Council for Mortgage Lenders reported that a recent inquiry into the problems (Coles, 1992) showed a marked correlation between possession orders and regional unemployment rates and stated that court orders had increased most rapidly in those regions where unemployment growth had been most rapid, where house prices had declined most sharply and where borrowers had higher levels of mortgage debt (and see Ford, 1994; Ford and Kempson, 1995). Moreover, whereas in the past arrears and possessions had been associated mainly with lower income, marginal owners, those getting into difficulties in the late 1980s recession included professional and managerial households. The Skipton Building Society reported in August 1994 that 44 per cent of its repossessions were professional borrowers - mainly doctors, architects and accountants.

By 1991 repossessions had risen to 75,000 and in 1992 households with arrears of 12 months or more topped 150,000. While these figures still represented a small proportion of all borrowers (for example, even at the

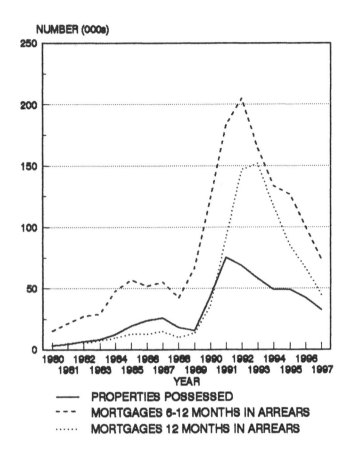

NUMBER (000s)

—— PROPERTIES POSSESSED
- - - MORTGAGES 6-12 MONTHS IN ARREARS
...... MORTGAGES 12 MONTHS IN ARREARS

Figure 2.4   Mortgage arrears and repossessions 1980-1997

Source: CML Market Briefing.

peak in 1991 repossessed properties accounted for 0.77 of all mortgaged properties), they were a major new source of misery for many households and an embarrassing policy problem for government and lenders. Moreover, the problem affected disproportionately those households which had bought at the height of the boom.   For example, a national survey showed that among those buying in 1988 some 7 per cent were in arrears in 1994/5 (Office for National Statistics, 1996).   A year later, while mortgage arrears had fallen slightly, some 1.4 million owners (one in 6) said that they were having some difficulty with mortgage payments (Office for National

16

Statistics, 1997a). Mortgage arrears also became a significant factor in homelessness, again with a strong regional dimension. In 1995 in England and Wales as a whole, mortgage arrears accounted for 8 per cent of all households accepted as homeless. In the East Midlands, the South East and the South West, however, the comparable figures were 15, 13 and 12 per cent respectively (Office for National Statistics, 1997b). Overall, in the period 1990 to 1995 some 350,000 properties were taken into possession compared with 129,000 in the whole of the previous decade.

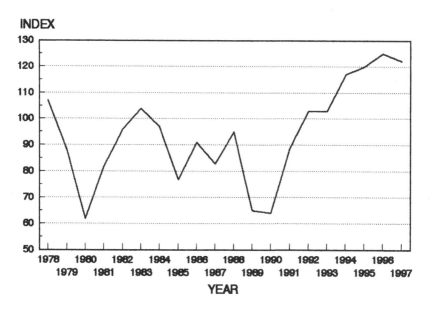

Figure 2.5   First time buyers' ability to buy index

Figures refer to first quarter of each year
Note: The higher the index, the greater the ability to buy

Source: National Housebuilding Council.

Concluding comments

Despite the depth and duration of the recession the majority of home owners emerged relatively unscathed from this unprecedented period of disruption and uncertainty in the housing market. Although home ownership grew substantially in the 1980s bringing in a wider variety of dwellings and households it was by then a mature tenure with a majority either owning outright or with diminishing and modest mortgage commitments. Nevertheless, it signalled a new climate of caution for the owner occupied market and the cohort which entered home ownership in the late 1980s and others trading up with limited equity experienced a very different set of market conditions than had previous generations of home owners. Financial deregulation, house price volatility, substantial fluctuations in interest rates, unemployment and job insecurity combined with other factors such as relationship breakdown to produce an unfamiliar environment for a tenure which had expanded in a period of rising prosperity and expanding job opportunities. Moreover, even among those home owners with steady employment and little or no mortgage debt, the equity in their property was exposed as being more vulnerable than they might have previously believed. This was a particular worry for the increasing number of home owners nearing or in retirement for whom the value of their dwelling was in most cases their major asset - albeit an asset of last resort to be drawn on if other resources became exhausted.

To what extent this period of change in the housing can be seen to have affected broader attitudes towards home ownership and housing market behaviour will be explored in greater detail in subsequent chapters. What is evident is that the rush to enter home ownership which characterised the housing market of the 1980s has not re-emerged despite house purchase being more affordable than at any time over the last twenty years (see figure 2.5). Home ownership is most attractive when house prices are rising steeply and it is least affordable. In calmer market conditions prospective owners are more inclined to wait rather than enter prematurely.

3 House price volatility and negative equity

The international context

The period of housing market instability outlined in the previous chapter was not confined to Britain. The growth and promotion of home ownership by governments has in recent decades been prevalent throughout most of the developed and developing world, though it has been pursued with varying strategies and degrees of enthusiasm (Munjee, 1995). The growth of home ownership has been accompanied by financial market deregulation, the internationalisation of capital markets and the integration of mortgage finance into global capital markets (Fallis, 1995). The stability of national housing finance systems evident in the 1970s has, to some extent, been undermined by these developments as housing finance has been exposed to global financial forces. According to Fallis 'The deregulation and shifts in world inflation rates led to the near collapse of the housing finance system in several countries' (Fallis, 1995, p.15). Bootle (1996) documents the volatility of house prices in the world's leading industrial countries. He argues that in every one of the G7 countries house prices rose 'remorselessly during the heyday of inflation' (Bootle, 1996, p.67). Between 1970 and 1992 the average annual rate of increase of house prices in Britain was 12.5 per cent, the same for Italy during 1970 to 1989. In the USA the annual rate was 7.75 per cent and in Germany 5.5 per cent. In real house price terms there was an average rise of 2 .5 per cent a year in Britain and Japan, 2 per cent in Canada and 1.5 per cent in Germany and the US (Bootle, 1996, p.68).

Kennedy and Andersen (1994) describe developments in house prices in fifteen industrialised countries between 1970 to 1992, asserting that 'in the majority of cases volatility appears to have increased in the 1980s. In real terms, Finland, Japan and the United Kingdom show the greatest volatility both for the period as a whole and for the 1980s (p.13) with the Netherlands following closely behind. They show the ten largest house price increases, and decreases, in any single year over the period 1970-92 (see Table 3.1). If we look particularly at developments occurring during the 1980s in terms of price increases in the top ten, Australia is ranked first with a peak of 38.1 per cent in 1988, Finland is second with 36.3 per cent in the same year and the United Kingdom is ranked third with 33.0 per cent, also in the same year. The sharpest decline was experienced by Finland in 1992, when prices fell by almost 17 per cent following a decline of almost 15 per cent in the previous year. Finland is followed by the Netherlands, then Norway, Sweden, Japan and Denmark.

Looking at what Kennedy and Andersen refer to as 'peak to trough movements', Finland heads the table with a 33 per cent fall in nominal house prices (40.3 per cent fall in real house prices) between 1989 to 1992. The United Kingdom is seventh with a 10.7 per cent fall in nominal house prices (23.0 per cent fall in real house prices) between 1989 to 1992.

These figures are useful as indicative averages but they inevitably mask local variation within countries. As Fallis, among others, has argued, 'there really are no *national* housing markets, rather there are *regional* (or local) housing markets (Fallis, 1995, p.6) which can perform in very different ways. He concentrates particularly on the United States, Britain, Australia and Canada agreeing with Bootle (1996) and Kennedy and Andersen (1994) that each of the countries experienced dramatic increases in house prices during the latter part of the 1980s. However, he emphasises the regional nature of these initial increases and subsequent volatility and points particularly to London in Britain, Boston in the United States, Sydney in Australia and Toronto in Canada. These areas experienced a dramatic escalation in house prices, linked to increased demand because of employment growth, immigration and income growth within the region, and an equally dramatic decline in prices in the early 1990s.

Table 3.2 highlights the housing market experiences of different regions in England and Wales between 1986 and 1996. It shows not only that nominal and real house price movement has been very different in northern and southern regions but also that the timing of peaks and troughs has varied across regions. It was the southern regions which led the upturn in house prices reaching their peak in 1989 and which have experienced the greatest percentage falls.

Whilst there is a recognition of and a growing literature about the international scope of housing market volatility there is very little published

Table 3.1

Boom and bust in nominal house prices: an international comparison

Rank		Country	Period	Magnitude*
Booms	1	United Kingdom	1972	48.2
	2	Netherlands	1977	39.7
	3	Australia	1988	38.1
	4	Finland	1988	36.3
	5	Japan	1973	34.7
	6	Japan	1972	33.3
	7	United Kingdom	1988	33.0
	8	Germany	1972	32.0
	9	United Kingdom	1979	30.1
	10	Netherlands	1976	28.6
Busts	1	Finland	1992	-16.9
	2	Finland	1991	-14.7
	3	Netherlands	1981	-10.3
	4	Netherlands	1982	-10.0
	5	Norway	1990	-9.3
	6	Sweden	1992	-9.2
	7	Japan	1974	-0.0
	8	Japan	1992	-8.7
	9	Netherlands	1980	-8.7
	10	Denmark	1987	-8.2

*Percentage change from previous year
Source: Kennedy and Andersen (1994).

material on negative equity. This is not an indication that negative equity is a peculiarly British phenomenon. The British context may appear unique given the combination of the availability of high loan-to-value ratio lending and the scale of nominal house price falls. However, Kennedy and Andersen (1994) show that the ratio of mortgage debt to the value of the owner-occupied stock of dwellings was generally higher at the end of 1992 than in previous years, and in addition to the UK, was particularly marked in Denmark, Finland, Canada and the United States, countries which have also experienced house price volatility.

Table 3.2
House price peaks and troughs by English regions and Wales, 1986-1996

Period	North	Yorks & Humber	East Mid.	East Anglia	South-East	South-West	West Mid.	North West	Wales
Peak	1994	1991	1990	1989	1989	1989	1990/1	1992	1990
Trough	1995	1994	1993	1993	1993	1993	1993	1993	1992
% Fall	5.1	4.6	7.2	22.0	19.4	20.6	4.2	3.9	2.4

Source: Table B.3 Compendium of Housing Finance Statistics, Council of Mortgage Lenders.

Note: The SE figure is an average of 16.7% fall in London and 22.7% fall in the rest of the South East. Note that nominal house prices in Scotland and Northern Ireland showed continuous rises throughout the 1986-95 period.

Source: Muellbauer and Cameron (1997) p.28.

In Denmark, where 52.8 per cent of the housing stock was owner-occupied in 1994, estimates for 1988 indicated that 25 per cent of home owners, particularly those under 40 years old, were 'technically insolvent' with net liabilities equivalent to 100-125 per cent of property value (10.2 per cent) and over 125 per cent (13.0 per cent) (Lunde, 1990). Kosonen (1995) and Timonen (1992) acknowledge the occurrence of negative equity in most Nordic countries but are particularly concerned about the rising incidence of housing debt problems amongst households in Finland. The number of households estimated to be in unmanageable housing debt problems in 1991 was between 20,000 and 30,000 and 25,000 in 1992 (Timonen, 1992 in Doling and Ruonavaara, 1996). For Canada and the USA the only evidence of negative equity appears to be anecdotal, with extensive searches in both countries failing to produce any relevant literature. There appears to be a more general concern with issues of affordability, arrears and repossessions. Negative equity is hypothetical and does not become a reality until a property is sold. It might therefore be seen as less of an immediate issue than mortgage non-payment, for example, by financial institutions and governments.

What is it and how many households are affected?

Holmans et al. (1996) describe negative equity as having a larger mortgage debt than the value of one's house or flat. For Thomas (1996) it involves comparing one asset (the individual's home) and one liability (the mortgage) only. Dorling et al. (p.2) expand the definition. They refer to it as

> a situation in which the (estimated) market price of a house has fallen below the original mortgage advance that was used to buy that house. Hence the owners find that the size of their mortgage is larger than the value of their home, and therefore that a portion of that mortgage is "unsecured" (Dorling et al., 1992, p.2).

There are a number of difficulties in calculating the extent and total value of negative equity. Numbers vary depending on methodology and data sources, values attributed to properties (estimated since 'real market value' can only be validated once competitive sale has taken place) and estimates of household debt levels. A broad discussion on the issues relating to definition and operationalising the concept of negative equity is offered by Dorling and Cornford (1995). Nevertheless, despite some criticisms (Cooper and Nye, 1995) the existing data do provide a broad picture of the scale of the problem and how it has evolved.

In 1992 the Bank of England estimated that a total of 876,000 households were in negative equity and calculated a total shortfall between outstanding loans and current valuations of £5.9 billion (Bank of England, 1992). More recent analysis indicated that the numbers in negative equity passed the million mark for the first time in 1993 and peaked in the first quarter of 1993 at 1.7 million, with the estimated aggregate value of negative equity at £10.8 billion (Council of Mortgage Lenders, 1996).

Holmans and Frosztega's (1996) study of negative equity analysed data from the 1992/93 and 1993/94 General Household Survey (GHS) to estimate the number of mortgaged owners who bought their homes in 1987 or later with negative equity. The study was therefore able to analyse information collected directly from owner occupiers. Studies carried out by the Bank of England (1992) and Wrigglesworth (1992) derived their estimates from the number of house purchase loans issued quarter by quarter, the distribution of ratios of mortgage advance to price, and changes in house prices since the quarter in which the advance was made. Thomas (1996) uses a similar method, including capital repayments on mortgages with Halifax house price data. He produced a time series calculation of the extent of negative equity. He estimated that in the first quarter of 1996 there were nearly one million households with negative equity with a total value of £4.0 billion, an average of £4,100 per household (Thomas, 1996). Table 3.3 using data from the Woolwich Building Society gives an historical and regional breakdown of negative equity estimates from 1992 to 1997 and shows clearly the numerical concentration in the south of England and also the different regional patterns of recession and recovery.

Table 3.3
Number of households with negative equity (000s)

	Greater London	Rest of South East	South West	North West	Other Regions
1992 Q4	240	550	190	80	510
1993 Q4	220	420	170	50	330
1994 Q4	170	380	130	70	375
1995 Q4	165	325	140	145	385
1996 Q4	30	135	75	25	170
1997 Q1	25	115	65	40	115

Source: Woolwich Building Society.

Dorling et al. (1992) study, in contrast, derived estimates from a sample of almost a million transactions between 1980 and 1991 by a single large building society instead of total loans by all lenders. This comprehensive study of the extent of negative equity concluded that one in ten of the 9,628,000 mortgage holders (CSO, 1992:154) 'had an unsecured portion of their mortgage at the start of 1992. They estimated that by the end of 1992 the proportion would be close to one in seven. In some regions of the country the figure is close to two out of every five of those who have borrowed since 1987' (Dorling et al., 1992, p.2).

From their study Holmans et al. (1996) were able to conclude that '769,000 households in Great Britain who bought in 1987 or later' had negative equity in 1993 (Holmans et al., p.6). Those with negative equity represented 3 per cent of 1987 purchasers, 31 per cent of 1990 purchasers and 8 per cent of 1993 purchasers. They went on to compare estimates from the General Household Survey (GHS) with the Family Expenditure Survey (FES). The FES indicated that of those households purchasing a property between 1987 and 1993, 906,000 were in negative equity. However, this initial estimate from the FES did not make the distinction between RTB purchasers and other purchasers as in the GHS. When adjusted the FES estimates of owner occupiers with negative equity who bought in 1987 or later was reduced to 730,000. Holmans et al. concluded that 'The GHS estimate of 769,000 households who bought in 1987 or later is consistent with the FES information'(Holmans et al., p.7). The authors also showed that these estimates were compatible with Bank of England estimates (adjusted for possessions) of 785,000 of mortgages homeowners who purchased in 1987 or later.

Similar calculations of mortgaged owner occupiers who bought their residences before 1987, give an estimated 90,000 (grossing the sample and eliminating RTB purchasers). Combining these figures Holmans et al. (1996) claim that the 'estimated number of owner-occupiers in Great Britain with negative equity is put at 860,000' (Holmans et al., p.8). It was also noted that all instances of negative equity among purchasers in 1987 identified in the GHS were due to remortgage and second mortgage. The same was likely to be true of those who bought in 1986 or earlier.

A related issue is that of insufficient equity, defined by Thomas (1996) as a situation in which a household 'possesses less than £5,000 worth of positive equity, which is generally insufficient to finance a move within the owner-occupied sector' (p.10). This category warrants important consideration given the uncertainties surrounding estimates of negative equity and the large number of households on the borders of positive/negative equity. There have been claims that in the first quarter of 1996, in addition to the estimated million households with negative equity, a further 2.1 million households suffered from insufficient equity. Thomas (1996) went on to estimate that the elimination of negative equity would require an increase in house prices in excess of 30 per cent. He forecast the extent of negative equity based on a five per cent increase in house prices between the final quarters of 1995 and 1996 and a six per cent increase between final quarters of 1996 and 1997. He calculated that by the end of 1997 265,000 households would be left in negative equity with an average equity shortfall of £3,400 and a total of £900 million (p.13). Figures from the Woolwich at the end of the first quarter of 1997 estimated that the number of households with negative equity fell by 45,000, reducing the numbers of households affected to 360,000, 'the lowest level since the first half of 1990' according to Earley (1997, p.8), and a sharp decrease from Thomas' estimate of 964,000 for the first quarter of 1996.

Who has negative equity and where is it?

There is general agreement that virtually all negative equity is held by households who have purchased between 1988 and 1991 (Gentle, Dorling and Cornford, 1994), Forrest et al. (1994; 1997). Holmans et al. (1996) confirmed date of purchase as a prime determinant of who experienced negative equity. The only study to consider the relationship between ethnicity and negative equity was our own in 1994 which concluded that 'non-white ethnic groups as a whole were significantly more likely to experience negative equity than white households' (p.2).

Negative equity has been recognised as a problem mainly affecting Southern England, particularly the South East. In their 1992 study Dorling et al. also pointed to this region as faring badly and went on to suggest that it was 'the poorer parts of the South East and South West which were most badly affected whereas those which had suffered least were generally in the poorer parts of Scotland, Wales and the North'. They claim that areas with the largest proportion of unsecured debt 'were those perceived as more "working class" in the south East'. It was also likely to be poor quality housing which was most affected (Dorling et al., 1992). Holmans et al. (1996) calculate that more than half (53 per cent including remortgages and second mortgages) of households with negative equity were in the South East with nearly three quarters, 73 per cent (76 per cent including remortgages and second mortgages) in southern England as a whole including East Anglia and the South West, a finding supported by Forrest et al. (1997).

Dorling and Cornford examined the distribution of negative equity between different categories of borrowers (Dorling and Cornford (1994) and in addition to the above categories, their study identified that negative equity was also likely to affect the following groups:

a)   those of lower incomes
b)   younger borrowers (under 25)
c)   those who purchased at the lower end of the market
d)   those in clerical or manual occupations.

They concluded

It is young, lowly paid, people without significant capacity to increase their incomes, and who have bought the cheapest kinds of housing, who are most likely to hold negative equity now, and who will still be most likely to hold negative equity in the near future, even after modest house price rises (1995, p.172).

We, however, reached a slightly different conclusion. Drawing on a social survey of over 1,200 households in Glasgow, Bristol and Luton as well as 30 in-depth interviews with households in negative equity in Bristol and Luton. Our findings showed that it was home owners with higher incomes rather than low incomes, and in 'social class A, and to a lesser extent social class B, [who] were much more likely than average to experience negative equity, while those in classes D and E, ... were not' (Forrest et al., 1994, p.9). In addition, the highest average negative equity (£10,167) was to be found amongst those in social class A. However, the largest proportion of people experiencing negative equity were in classes B, C1 and C2. A later study concentrated on new build estates in Bristol and Luton specifically because it was felt that 'the new build estates of the late 1980s had been disproportionately affected by the recession in the residential property market' (Forrest et al., 1997, p.2). These were areas which had experienced some of the steepest falls in property values in the early 1990s and contained concentrations of first time buyers, households which had moved during a period of escalating house prices and a large number of smaller starter homes.

First-time buyers have been identified as being disproportionately affected by negative equity (Dorling et al., 1995 and Forrest et al., 1994). Holmans et al. (1996) also showed that this was an important factor in that first time buyers were three times more likely to suffer from negative equity than previous owners (though they do go on to say that the mean amounts of negative equity were much higher among moving owner-occupiers than among first time buyers (p.16), a similar finding to that in Forrest et al. (1994). However, there is disagreement with Dorling et al.'s claim that those with negative equity were 'virtually all first-time buyers' and under 25. We found that one-third of our sample had previously owned another property. And whilst Holman et al. (1996) recognise that the risk of having negative equity is highest among young households as they 'have a shorter time in which to accumulate a deposit' (p.17) they argue that this does not mean that most households with negative equity are young. Around a third of heads of household aged under 25 had negative equity with their mean negative equity being £2,900. However, young householders (under 25) form a relatively small proportion of all mortgage holders (6 per cent). Just over 80 per cent of heads of household with negative equity are aged between 25 and 44 (an age group incorporating 74 per cent of all mortgage holders), with 7 per cent 45 and over.

Whilst Dorling et al. (1992) argued that social class was the most significant indicator of the households most likely to experience negative equity, we concluded that negative equity 'is not a problem of "marginal" homeowners' (Forrest et al., 1994 p.2) but has affected a wider cross-section of home owners. Holmans et al. also show that the percentage distributions of all mortgage holders and of mortgage holders with negative equity are very similar 'suggesting that negative equity has affected all social groups' (Holmans, 1996, p.18).

Future prospects

Over the last few years there has been a flurry of research activity around the phenomenon of negative equity. A number of apparently false starts to housing market recovery prolonged the agony for many home owners and offered the prospect of a home ownership market which had changed in fundamental ways. Gloomier predictions of entrenched unsecured housing debt were revised with evidence of a strong if uneven price recovery. In the 12 month period from September 1995 the Council for Mortgage Lenders reported a modest increase in transactions, an increase in net advances and house price rises of 5-7 per cent (Council for Mortgage Lenders, 1996). Earley (1997) demonstrated this continuing trend with data from Halifax and Nationwide house price indices showing prices continuing to increase at 7.2 per cent and 9.6 per cent year on year respectively. In the first quarter of 1997 prices rose by 1.3 per cent, 8.6 per cent up on the previous year (Nationwide, 1997).

Regional house price changes are diverse and in 1997 The Halifax index showed a range of between 1.8 per cent in the North and 17.4 per cent in Greater London with 'little evidence of across the board house price increases in particular areas' (Earley, 1997, p.6). This supports our own analysis (Forrest et al., 1994) that an uneven recovery would see some parts of the housing market lagging behind, particularly small starter homes and studio flats. Whilst modest house prices might lift the majority of borrowers out of negative equity a rump of properties and people with the most intractable problems may be left behind. Nevertheless, there is clear evidence of a substantial decline in the number of households in negative equity. Figure 3.1 uses Woolwich Building Society data for the UK as a whole and shows that the number of households affected by negative equity fell from a peak of 1.7 million in the first quarter of 1993 to 360,000 in the first quarter of 1997 – equivalent to levels in the early 1990s.

28

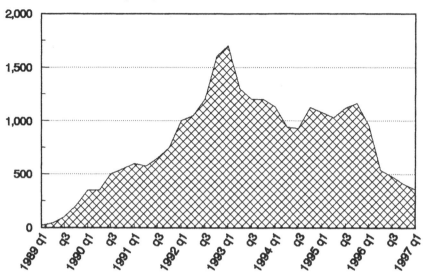

HOUSEHOLDS (000s)

Figure 3.1   The rise and fall of negative equity, UK 1989-1997

Source: Department of Employment, Transport and the Regions.

According to Bootle (1996) the peoples 'love affair with property' and their preoccupation of 'buying early, borrowing as much as possible, moving to larger and larger properties' (p.66) seems to be diminishing in what he refers to as the zero, post-inflationary era. Other studies have suggested that there has been a loss of confidence in home ownership as an investment (Maclennan, 1994) and greater concern amongst current or prospective owners towards mortgage indebtedness (DoE, 1994: Office for National Statistics, 1996). However, evidence is contradictory making it difficult to ascertain whether there has been any fundamental change in attitudes or behaviour in relation to home ownership. In the Housing Attitudes Survey (DoE, 1994), for example, 87 per cent of households agreed that property was a good investment (p.64) following the most severe house price fall in living memory.

Kempson and Ford's (1995) study of attitudes and beliefs towards home ownership and borrowing shows that over the past twenty years there has been an increasingly positive attitude to home ownership, with a marked increase in the proportion of adults saying they would most like to own their own home in the next ten years (from 62 per cent in 1975 to 85 per cent in

1993). The same trend is true for adults expressing a preference for owner occupation in the short term, but with a slight reduction for the years 1991 and 1992. Using data from BSA/CML surveys (BMRB 1991, 1993) Kempson and Ford (1995) also considered whether the attractiveness of owner occupation had changed over the past two years. In 1991 and 1993 more people said they considered home ownership more attractive (35 per cent and 42 per cent) than less so (20 per cent and 28 per cent). In 1992, however, 33 per cent of respondents considered owner occupation less attractive, with only 21 per cent considering it more so. For Kempson and Ford (1996) these findings suggest that 'beliefs about home ownership took a temporary knock at a time when problems in the housing market were receiving a good deal of coverage in the media - mortgage arrears and possessions were at their peak and there were large falls in house prices' (p.21).

4 Home owners in negative equity

Introduction

None of the national estimates of the prevalence of negative or reduced equity discussed in the previous chapter include significant information on the characteristics of the households which are affected or the dwellings which they occupy. Studies based on information from mortgage lenders in particular suffer from the disadvantage that even the limited household data which are available relate to the point at which the mortgage commenced and is likely to have changed to a greater or lesser extent in the subsequent period. Furthermore our own work for the Joseph Rowntree Foundation, although far more detailed in terms of information on those experiencing negative equity, is based on household samples in only three cities - Glasgow, Bristol, and Luton, and cannot in any sense be argued to provide a representative picture of the whole population of home-owners in negative equity. Our subsequent in-depth follow-up work (discussed in the following chapters) was not intended to provide such a picture but instead sought to explore attitudes to negative equity and behavioural responses.

However, one data source provides a possibly unique picture of the characteristics of those experiencing negative and low equity - the English House Condition Survey. This quinquennial study, a complex set of inter-linked surveys rather than a single one, is commissioned and co-ordinated by the Department of the Environment, Transport and the Regions, and analysed by the Department and by staff from the Building Research Establishment (see Department of the Environment 1993 for a full description and for basic results). Although the main focus of the survey is the measurement of physical dwelling conditions in a sample of around 25,000 dwellings, it also includes a linked questionnaire survey of a sample of around 10,000 households living at the sampled addresses, a survey of information held by local authorities on these dwellings, and the provision of professional valuations of sampled properties. The combination of independent data on property values, together with detailed information on outstanding mortgage debt, enables the English House Condition Survey to be used to derive estimates of the home equity holdings of owners in the sample. These can then be related to a wide range of household and dwelling characteristics and to information on financial commitments, investment behaviour in relation to repair, maintenance and improvement, and attitudes towards the dwelling and the surrounding area. Our study of negative and low equity for the Department of the Environment, Transport and the Regions included the provision of access to previously unpublished data from the 1991 and 1996 English House Condition Surveys and this chapter describes and expands on the findings to present a broader picture of the range of circumstances of households with negative and eroded equity in 1991.[1]

Professional and self valuation

The availability of professional valuation data enabled this analysis to draw on objective estimates of the amount of equity held by home-owners in the sample, as distinct from owners' own estimates of their equity position which form the basis for the analysis in Chapters 5-7. However, since the 1991 English House Condition Survey provides both independent and home owner valuations, it is possible to compare the two (Table 4.1). Some 51 per cent of owners in 1991 produced an estimate of value more than 10 per cent higher than a professional valuer, with 11 per cent estimating their house was worth over 50 per cent more than the valuer. For those with negative equity (assessed on the basis of the independent valuation, not their own) three quarters were more optimistic than the independent valuer, with a fifth estimating their home to be worth over 50 per cent more than the valuer's figure. This clearly shows a tendency for households to avoid admitting that they were in negative equity by maintaining an inflated estimate of house value.

Those with low positive equity were more likely than average to produce an estimate of value close to that of the professional valuer, while those with equity of £5,000 or more were more likely than average to under-estimate their equity, although even then only 12 per cent did so in practice.

A further indication of the impact of negative equity is provided by the high number of home-owners who produced self valuations which equated

32

exactly with their outstanding mortgage. In contrast, the independent valuations produced relatively few such cases except amongst very recent movers. This suggests that many owners whilst unwilling to claim positive equity were also unwilling to admit to negative equity.

Table 4.1
Comparison of owner's estimate of house value with professional valuation by amount of equity held[2]

Estimated home equity in 1991

Comparison of owner's and professional valuation of dwellings	Negative or none	Positive up to £5,000	Positive over £5,000	All owner occupiers
Owner over 50 per cent in excess of valuer	19.7	12.9	10.9	11.3
Owner 11-50 per cent in excess of valuer	53.4	37.9	39.4	39.8
Owner within 10 per cent of valuer	24.4	46.4	36.8	36.6
Owner 11-50 per cent less than valuer	2.1	2.9	12.1	11.5
Owner over 50 per cent less than valuer	0.5	0.0	0.8	0.8
All owners	100.0	100.0	100.0	100.0
Base	193	140	4807	5140

Source: 1991 English House Condition Survey, our analysis.

Negative and low equity overall

Throughout the following analysis, comparisons will be made between owner occupiers in negative equity, those having low equity (up to £5,000), other home owners aged under 45, and owners with over £5,000 in equity aged 45 or more. Those with positive equity in excess of £5,000 have been separated out into younger and older households order to isolate the large group of older home owners mostly owning their homes outright.

Table 4.2 shows numbers of weighted and unweighted cases in each equity status category. Subsequent tables are based on weighted data which are grossed to national totals. The 1991 English House Condition Survey estimated that some 454,000 owner occupiers were in negative equity (3.5 per cent of all owners), with a further 322,000 (2.5 per cent) having positive equity of up to £5,000.

Table 4.2
1991 English house condition survey sample

Equity status	Unweighted No	Weighted No (000s)	Per cent
Negative or no equity	202	454	3.5
Positive equity under £5,000	147	322	2.5
Positive equity over £5,000, aged under 45	1,928	4,363	33.9
Positive equity over £5,000, aged 45 or more	3,258	7,334	60.1
All owner occupiers	5,535	12,872	100.0

Source: 1991 English House Condition Survey, our analysis.

Purchase characteristics

Table 4.3 shows selected characteristics of the household's most recent house purchase by equity status in 1991. A purchase date for the current dwelling falling between 1988 and 1991 was strongly associated with negative equity. Some 89 per cent of those in negative equity and 84 per cent of those with low positive equity had purchased their dwelling in the four years prior to the survey (1988-91), compared with only 37 per cent of younger and 13 per cent of older households with over £5,000 of equity. Of current owners who had purchased in the 1988-91 period 12 per cent were in negative equity compared with 0.5 per cent of those who had bought in earlier years.

Those with negative equity were also more likely to be first-time buyers (43 per cent) than households with low positive equity (37 per cent) or those with over £5,000 of equity, especially if home-owners who purchased their current dwelling from a local authority or housing association are excluded.

Table 4.3
Purchase characteristics by equity status,
owner occupiers in England 1991

	Negative or none	Positive up to £5,000	Positive over £5,000, aged under 45	Positive over £5,000, aged 45 or more	All home owners
Date of purchase					
Pre 1987	6.0	13.0	52.1	80.6	66.6
1987	5.3	6.5	11.2	6.0	7.7
1988	27.6	8.3	11.7	4.7	8.0
1989	13.9	23.1	8.8	3.5	6.2
1990	19.6	23.1	8.4	2.6	5.7
1991	27.6	25.9	7.8	2.6	5.8
First time buyer including from LA/HA					
Yes	43.0	37.3	34.7	35.4	35.5
No	57.0	62.7	65.2	64.3	64.3
First time buyer excluding from LA/HA					
Yes	42.6	37.3	30.9	29.3	30.6
No	57.4	62.7	69.0	70.4	69.2
Original mortgage as per cent of purchase price, all owners					
Up to 80 per cent	18.5	17.4	57.0	37.1	42.7
81-90 per cent	15.2	33.2	10.6	8.3	9.9
90 per cent or more	66.2	49.4	32.4	54.6	47.4
Original mortgage as per cent of purchase price, bought in last four years					
Up to 80 per cent	13.3	15.9	57.5	32.7	41.7
81-90 per cent	16.5	32.6	9.2	2.6	9.7
90 per cent or more	70.2	51.5	33.3	64.7	48.6

Source: 1991 English House Condition Survey, our analysis.

However, this still implies that a majority of those found to be in negative equity in the 1991 English House Condition Survey were **not** first time buyers. It should also be noted that only 4.3 per cent of all first-time buyers and 5.4 per cent of first time buyers purchasing from a private owner were in negative equity, not much more than the 3.5 per cent of owners as a whole in this position. Amongst first time buyers who had bought in the previous four years, however, some 24.4 per cent were in negative equity and 13.1 per cent had low positive equity, compared with only 7.9 per cent and 5.8 per cent of previous owners. This corresponds with the finding from our various in-depth studies which suggested that many of those in negative equity were owners who had leaked equity from a previous purchase to purchase furniture or a car or for some other purpose, or whose gains on their earlier purchase had simply been swallowed up by the fall in the value of their current property.

A high percentage mortgage was also associated with negative equity, with two thirds of those in negative equity having taken out a mortgage of 90 per cent of purchase price or more, rising to 70 per cent if those resident for over four years are excluded. Only a half of those with low positive equity and under a third of younger households with equity of over £5,000 had a 90 per cent mortgage or higher.

Household characteristics

There were clear differences in household characteristics between those with negative or low equity and other home-owners (Table 4.4). Those with negative equity were typically younger, with 41 per cent headed by a person under 30, compared to only 8 per cent of owner occupiers as a whole. Only 9 per cent of households with negative equity were headed by a person aged 45 or more. Just under one in five households headed by a person aged under 30 had negative equity, about four times the proportion for those above this age.

Given this age profile for the household head, households with negative equity were not surprisingly also more likely to be single adults, and couples without children, or more generally people at early stages in the life cycle. It should be recalled that the data relate to the situation in 1991, since when some households in these categories may have formed partnerships and/or had children. For the same reason they were more likely to be single or cohabiting and less likely to be married or widowed. Households headed by a member of a minority ethnic group were very slightly more likely to be in negative equity than white-headed households.

36

Table 4.4
Household characteristics by equity status,
owner occupiers in England 1991

| | Percentage by equity status category | | | | |
	Negative or none	Positive up to £5,000	Positive over £5,000, aged under 45	Positive over £5,000, aged 45 or more	All home owners
Age of household head					
16-29	41.2	32.5	15.6		7.6
30-44	49.8	55.4	84.4		31.7
45-59/64	9.0	11.1		55.4	33.9
60/65-75	0.0	0.9		27.3	16.4
75 or older	0.0	0.0		17.3	10.4
Household type					
Lone adult under 60/65	12.8	12.7	7.2	5.3	6.4
Two adults under 60/65	49.2	39.1	17.5	18.0	19.4
Lone parent	0.0	2.8	5.3	0.4	2.1
Small family	23.0	32.6	43.1	5.2	19.4
Large family	10.2	8.1	21.3	6.6	11.7
3 or more adults	4.9	4.7	5.5	18.4	13.2
Two adults at least one 60/65 or older	0.0	0.0	0.2	27.5	16.6
Lone adult 60/65 or older	0.0	0.0	0.0	18.6	11.2
Gender of household head					
Male	89.9	91.0	89.6	80.1	84.0
Female	10.1	9.0	10.4	19.9	16.0
Marital status					
Married	57.3	61.9	75.6	69.0	70.6
Cohabiting	23.3	19.5	7.5	1.1	4.5
Single	11.2	11.1	7.4	5.1	6.2
Widowed	0.0	0.0	0.9	18.7	11.6
Separated/divorced	8.1	7.5	8.5	6.1	7.0
Ethnic group					
White	95.1	95.3	95.0	97.2	96.3
All others	2.0	1.2	1.5	1.2	1.3
Base	454	323	4363	7734	12874

Source: 1991 English House Condition Survey, our analysis.

Overall the table confirms the picture of negative equity as mainly affecting younger households at the earlier stages of their housing careers.

Economic status

Households with negative equity in 1991 were more affluent than owner occupiers with positive equity (Table 4.5), but this is not surprising given the substantial group of low income older home-owners. However people with negative equity were also more likely to have a higher net household income than younger households with positive equity. Some 47 per cent of households with negative equity had a net household income of £20,000 per annum or more compared with 39 per cent of those with up to £5,000 in positive equity and 30 per cent of other owners aged under 45. This reflects the fact that to be able to purchase at all in the 1988-91 house price boom period, when the risk of going into negative equity was greatest, required a substantial income.

Those with negative equity were also predominantly from professional and non-manual socio-economic groups. Some 75 per cent of those with negative equity fell into these categories, compared with 58 per cent for those under 45 with over £5,000 in equity. Interestingly, a high proportion of those with low positive equity were from the semi-skilled manual socio-economic group. Only 6 per cent of household heads with negative equity were not in employment, but this was higher than for those with positive equity, irrespective of age.

More than half of households in negative equity had two full time earners (53 per cent) and only 7 per cent had no earner. The profile for those with low positive equity was similar. Younger households with over £5,000 in equity who were longer established in the housing market were typically much less likely to have two full-time earners (only 25 per cent did so), probably because they were more likely by this stage to have children. Not surprisingly, over 50 per cent of home-owner households over 44 with positive equity had no earners because a high proportion were retired.

Overall, the profile of older home-owners with over £5,000 in equity differs from younger owners on all economic indicators because of the high proportion who are retired. Those with negative equity are typically younger double earner households with higher incomes, predominantly from professional and non-manual socio-economic groups. Some of those with low positive equity display similar trends, but this group also includes some people more well-established in the housing market who simply occupy lower value properties.

38

Table 4.5
Economic characteristics by equity status,
owner occupiers in England 1991

	Percentage by equity status category				
	Negative or none	Positive up to £5,000	Positive over £5,000, aged under 45	Positive over £5,000, aged 45 or more	All home owners
Net household income (£ per annum)					
Under £8000	9.5	6.5	10.3	43.4	30.0
£8000-11999	7.3	13.7	14.0	18.3	16.4
£12000-15999	11.7	15.3	23.7	14.4	17.5
£16000-19999	24.7	25.9	21.3	10.4	15.0
£20000-23999	18.1	14.6	11.8	5.3	8.2
£24000 or more	28.9	24.0	18.9	8.2	12.9
Socio-economic group of household head					
Professional and managerial	40.9	31.0	37.1	34.5	35.5
Intermediate and junior non-manual	34.3	22.3	21.2	22.1	22.2
Skilled manual	21.1	30.3	31.1	27.5	28.6
Semi-skilled manual	2.2	13.6	7.0	9.7	8.6
Unskilled and other	1.5	2.8	3.5	6.2	5.0
Employment status of household head					
Employed full-time	88.1	92.0	88.9	39.5	59.2
Employed part-time	2.0	1.5	3.2	5.0	4.2
Unemployed	6.0	2.8	4.5	3.7	4.0
Retired	0.0	0.9	0.2	46.3	27.9
Other	4.0	2.8	3.2	5.5	4.6
Number of earners					
Two full-time	52.6	52.2	24.4	10.8	17.9
At least one full-time	37.0	39.1	64.5	26.7	40.2
One or more part-time	3.7	3.1	5.5	8.8	7.4
None	6.6	5.6	5.2	53.5	34.3
Base	454	321	4360	7734	12869

Source: 1991 English House Condition Survey, our analysis.

Dwelling characteristics

Table 4.6 shows some basic characteristics of the dwellings occupied by households with negative and low equity in comparison to other home-owners. Households in negative equity and with low positive equity were over-represented in pre-1919 and terraced dwellings, and in converted or purpose-built flats. Not surprisingly, they were also over-represented in smaller dwellings and those with lower values. Some 44 per cent of those in negative equity in 1991 and 31 per cent of those with low positive equity lived in pre-1919 dwellings, compared to around 28 per cent of owner occupiers with greater amounts of equity. Negative equity in 1991 was thus very much a phenomenon of the older housing stock. Almost half of those with negative equity lived in properties with a value of under £50,000 and 81 per cent lived in properties valued at less than £73,000, the average for 1991 in England. The average property value for those with negative or low equity was some £15-20,000 lower than for younger households with more than £5,000 in equity.

Dwelling conditions and investment in repair and improvement

In terms of dwelling condition indicators (Table 4.7), those with negative equity were the least likely to live in housing which was deemed to be unfit for human habitation and those with low positive equity were the most likely to do so, although it must be stressed that with this indicator, numbers were very small. In contrast, for disrepair, households in negative equity were less likely to have nil or very low outstanding repair costs and considerably more likely to have repair costs in the £1,000-£5,000 band, but less likely to have costs above this level. Those with over £5,000 in equity generally faced lower repair costs than those with negative or low positive equity. In terms of overall averages these differences cancelled one another out. Looking at repair costs per square metre, in order to take account of dwelling size, those with negative equity were less likely to live in the 30 per cent of dwellings with the lowest repair costs, but there was no strong pattern. This suggests that negative equity is not strongly associated with poor physical conditions as measured by these indicators.

Looking at households' own assessment of the state of repair of their dwelling, however, there was considerably less satisfaction with the state of repair of the dwelling for those in negative equity and with low positive equity than for those with £5,000 or more. Since this cannot be explained by the objective indicators of physical condition discussed above, it suggests

Table 4.6
Dwelling characteristics by equity status,
owner occupiers in England 1991

	Negative or none	Positive up to £5,000	Positive over £5,000, aged under 45	Positive over £5,000, aged 45 or more	All home owners
	\multicolumn{5}{c}{Percentage by equity status category}				
Age of dwelling					
Pre 1919	44.1	31.1	28.1	27.2	28.2
1919-1944	15.6	18.0	20.6	20.7	20.4
1945-1964	12.3	12.7	15.7	21.5	19.0
Post 1964	28.0	38.2	35.6	30.6	32.4
Type of dwelling					
Terraced	41.3	63.9	30.1	23.6	27.4
Semi-detached	17.9	21.9	37.8	34.0	34.4
Detached	11.7	5.3	26.6	32.3	29.0
Converted flat	11.5	1.3	2.0	6.0	4.7
Purpose built flat	17.7	7.5	3.5	4.1	4.5
Size of dwelling					
Very small (under 50 square metre)	20.0	23.7	7.7	7.3	8.3
Small (under 50-69 square metre)	39.2	42.4	30.1	28.8	29.9
Medium (70-99 square metre)	28.0	21.8	37.3	34.2	34.7
Large (100-149 square metre)	9.9	10.3	19.1	18.8	18.4
Very large (150 square metre or more)	2.9	1.9	5.8	11.0	8.7
Value in 1991					
Under £25,000	4.6	2.5	0.9	2.5	2.1
£25,000-£49,000	40.8	48.1	27.2	25.0	26.9
£50,000-£74,999	37.7	38.2	40.3	36.6	38.0
£75,000-£99,999	12.1	9.3	16.0	16.1	15.7
£100,000-£149,000	4.6	1.9	11.3	12.4	11.5
£150,000 or more	0.0	0.0	4.2	7.3	5.8
Mean dwelling value 1991					
£	55,945	52,433	70,678	76,226	73,036
Base	454	322	4363	7734	12873

Source: 1991 English House Condition Survey, our analysis.

41

that negative and low positive equity circumstances are a factor leading to negative attitudes, an issue discussed further below.

There is also no evidence to suggest that those with negative equity or low positive equity were either more or less likely to invest in their homes than other households (Table 4.7). Older households with over £5,000 in equity and those with low positive equity were the most likely groups to have invested in repair, maintenance and improvement in 1991. However in terms of the amount spent, those with low positive equity had the highest average investment, perhaps because they had borrowed to finance work, while older households with £5,000 or more in equity had on average spent the least. In terms of the type of spending, there were again no clear trends although those with negative equity tended to have spent a slightly higher proportion of their total investment, on average, on improvement rather than on repairs.

Attitudes to dwelling and neighbourhood

The 1991 English House Condition Survey provides a wealth of information on levels of satisfaction with various aspects of the home, the immediate area around the dwelling, and the wider neighbourhood. How does housing equity status affect various aspects of satisfaction with the dwelling and neighbourhood? Table 4.8 shows overall attitudes towards the dwelling, the surrounding area and the neighbourhood, together with views on some detailed aspects of the dwelling itself.

In general, negative and low positive equity households were less likely to be satisfied with their home overall, and with the number of rooms, than younger households with £5,000 or more in equity, although views on the size of rooms did not differ by much from the overall picture. Older households with £5,000 or more in equity were the most likely to be satisfied on all the aspects examined in the table. Our previous work on negative equity (Forrest, Kennett and Leather, 1994, 1997) has shown that households in negative equity were often dissatisfied with their homes because they were unable to move to a larger dwelling (following the birth of children, for example). This also emerges from the 1991 English House Condition Survey in terms higher levels of dissatisfaction with the number of rooms available to the household.

Households in negative equity or with low positive equity were also less satisfied than other owners with their surroundings and slightly less satisfied with the neighbourhood, again consistent with our findings in previous studies. However they were more likely to be undecided than dissatisfied,

Table 4.7
Dwelling condition and investment in repair and improvements by equity status, owner occupiers in England 1991

	Percentage by equity status category				
	Negative or none	Positive up to £5,000	Positive over £5,000, aged under 45	Positive over £5,000, aged 45 or more	All home owners
Unfitness					
Fit	96.3	92.2	95.4	94.3	94.7
Unfit	3.7	7.8	4.6	5.7	5.3
General repair costs					
None	22.9	23.5	24.0	24.6	24.3
Under £1,000	39.9	44.0	44.8	44.4	44.4
£1,000-£4,999	35.0	26.3	28.4	26.6	27.5
£5,000 or more	2.2	6.2	2.8	4.3	3.8
Average repair costs (£)	1,100	1,167	1,012	1,132	1,091
Repair costs per square metre					
Best 30 per cent of dwellings	28.5	37.3	32.7	33.9	33.4
Rest	63.1	47.5	60.2	58.9	59.2
Worst 10 per cent of dwellings	8.4	15.2	7.1	7.2	7.4
Satisfaction with state of repair					
Satisfied	49.6	59.6	61.5	74.8	69.0
Neither satisfied nor dissatisfied	41.2	26.4	27.3	18.4	22.4
Dissatisfied	9.3	14.0	11.2	6.8	8.6
Work carried out in 1991					
Yes	16.8	18.9	13.7	18.1	16.6
No	83.2	81.1	86.3	81.9	83.4
Value of work carried out in 1991					
None	16.8	18.9	13.7	18.1	16.6
Under £1,000	27.2	27.2	36.2	41.5	38.8
£1,000-£4,999	41.9	32.5	34.0	31.8	32.9
£5,000 or more	14.1	21.4	16.1	8.6	11.7
Average value of work (£)	2,679	3,580	2,677	1,795	2,170
Per cent of 1991 expenditure on dwelling devoted to repairs					
Up to 25 per cent	54.5	53.1	53.8	52.2	52.8
26-50 per cent	26.2	19.8	22.6	16.9	19.3
51-75 per cent	9.3	7.6	10.7	10.8	10.7
Over 75 per cent	10.1	19.5	12.9	20.1	17.2
Base	454	322	4363	7734	12873

Source: 1991 English House Condition Survey, our analysis.

43

Table 4.8
Attitudes to home, area and neighbourhood by equity status,
owner occupiers in England 1991

	Percentage by equity status category				
	Negative or none	Positive up to £5,000	Positive over £5,000, aged under 45	Positive over £5,000, aged 45 or more	All home owners
Satisfaction with home					
Satisfied	83.0	86.9	88.4	92.7	90.7
Neither satisfied nor dissatisfied	14.3	8.7	7.5	5.6	6.6
Dissatisfied	2.6	4.4	4.1	1.8	2.7
Satisfaction with no. of rooms					
Satisfied	68.7	70.2	64.7	84.7	77.0
Neither satisfied nor dissatisfied	12.8	17.7	17.1	9.8	12.5
Dissatisfied	18.5	12.1	18.2	5.6	10.5
Satisfaction with size of rooms					
Satisfied	76.8	75.2	71.6	83.0	78.8
Neither satisfied nor dissatisfied	14.8	17.4	19.0	12.4	14.9
Dissatisfied	8.4	7.5	9.4	4.5	6.4
Satisfaction with surroundings					
Satisfied	78.4	82.9	83.7	88.1	86.1
Neither satisfied nor dissatisfied	15.2	11.2	11.4	7.9	9.4
Dissatisfied	6.4	5.9	4.7	3.7	4.2
Satisfaction with neighbourhood					
Satisfied	82.6	85.7	83.6	88.3	86.5
Neither satisfied nor dissatisfied	14.5	12.4	12.7	9.0	10.6
Dissatisfied	2.9	1.9	3.5	2.5	2.8
Base	454	322	4363	7734	12873

Source: 1991 English House Condition Survey, our analysis.

suggesting that they did not feel so strongly about these aspects as they did about features of the dwelling itself.

Attitudes to repair, maintenance and improvement

Table 4.9 examines attitudes to repair, maintenance and improvement by equity status. Respondents were asked to say whether they agreed or disagreed with a range of statements relating to repair, maintenance and improvement. Responses have been aggregated to show the percentages agreeing with each statement. Few owners in any equity status group accepted that they were not interested in the condition of their homes. Those with negative equity appeared less likely to regularly inspect their homes to see if repairs need action and were less likely than other owners to feel that they took a preventative approach to repair work. They were more likely to feel that repairs cost more than they could afford - for example 44 per cent of those with negative equity agreed with this statement, compared with 32 per cent of older households with over £5,000 in equity. Yet as Table 4.5 above showed, those with negative equity were in practice more likely than others to have an income adequate to cope with repair costs. Those with negative equity were more likely to agree that repairs did not add to value (22 per cent) or to profess ignorance on which types of investment added most to house value (16 per cent). They were also less likely to agree that improvements soon pay for themselves and less likely to enjoy home improvements. These findings are sufficiently consistent to suggest that those with negative equity do have less positive attitudes towards repair, maintenance and improvement than those with more equity. However this was only relative - large majorities of those in negative equity still felt that repairs and improvements could increase values or pay for themselves in a reasonable time. There was no evidence of **strongly** negative attitudes towards the home or its surroundings amongst those with negative equity.

Moving intentions

Table 4.10 shows intention to move in the next five years by equity status. Those with negative equity were very much more likely to be seeking a move in the next five years (57 per cent) than any other group. A significant proportion of households with low positive equity (47 per cent) were also seeking to move, whilst around 40 per cent of younger households with positive equity in excess of £5,000 wish to move. A much lower proportion

Table 4.9
Attitudes to repair, maintenance and improvement by equity status, owner occupiers in England 1991

	Percentage by equity status category				
	Negative or none	Positive up to £5,000	Positive over £5,000, aged under 45	Positive over £5,000, aged 45 or more	All home owners
Regularly inspect for work					
Agree	44.1	63.2	52.5	59.3	56.5
Disagree or no opinion	55.9	36.8	47.5	40.7	43.5
Not interested in condition of home					
Agree	5.9	4.3	7.4	8.6	8.0
Disagree or no opinion	94.1	95.7	92.6	91.4	92.0
Repair to avoid future problems					
Agree	78.6	75.2	82.3	83.9	83.0
Disagree or no opinion	21.4	24.8	17.7	16.1	17.0
Only do repairs in emergency					
Agree	13.7	9.6	11.5	18.4	15.7
Disagree or no opinion	86.3	90.4	88.5	81.6	84.3
Repairs cost more than can afford					
Agree	44.1	33.2	31.9	32.2	32.5
Disagree or no opinion	55.9	66.8	68.1	67.8	67.5
Repairs don't add to value					
Agree	22.0	10.9	13.5	16.2	15.4
Disagree or no opinion	78.0	89.1	86.5	83.8	84.6
Don't know which improvements add to value					
Agree	15.9	6.2	9.7	16.9	14.2
Disagree or no opinion	84.1	93.8	90.3	83.1	85.8
Improvements soon pay for themselves					
Agree	58.5	63.6	61.5	65.2	63.7
Disagree or no opinion	41.5	36.4	38.5	34.8	36.3
Enjoy improving home					
Agree	69.5	84.2	74.8	72.5	73.5
Disagree or no opinion	30.5	15.8	25.2	27.5	26.5

Source: 1991 English House Condition Survey, our analysis.

Table 4.10

Moving intentions by equity status, owner occupiers in England 1991

	Percentage by equity status category				
	Negative or none	Positive up to £5,000	Positive over £5,000, aged under 45	Positive over £5,000, aged 45 or more	All home owners
Intention to move in next five years					
Yes	57.0	47.2	39.6	18.3	27.6
No	33.7	46.6	53.8	73.1	64.5
Don't know	9.3	6.2	6.5	8.5	7.8
Base	454	322	4363	7335	12874
Reason for wanting to move					
Another job	16.2	23.0	15.7	5.0	11.9
More suitable area	18.8	18.5	21.8	15.9	19.1
More suitable/better house	64.0	65.6	61.0	45.9	55.5
Family/personal reasons	26.1	34.8	17.8	35.6	26.1
Can't afford to live here	4.4	0.0	2.4	4.9	3.4
Other	13.6	8.6	16.1	28.7	20.5
Base	259	152	1718	1370	3599
Actions taken to move					
Nothing	39.8	54.8	37.2	39.5	39.1
Looked at rent levels	3.1	0.0	1.5	1.7	1.6
Looked at purchase prices	38.9	34.5	44.9	38.3	41.4
Viewed houses	16.7	30.5	25.0	23.1	23.9
Started saving money	15.2	12.0	12.1	3.0	8.7
Put house on market	25.1	22.4	26.2	26.1	25.9
Bought another house	2.5	0.5	3.0	1.2	2.2
Base	259	150	1724	1414	3456

Source: 1991 English House Condition Survey, our analysis.

47

Table 4.11

Regional variations in equity status, owner occupiers in England 1991

	Percentage by equity status category				
	Negative or none	Positive up to £5,000	Positive over £5,000, aged under 45	Positive over £5,000, aged 45 or more	All home owners
Northern	2.2	4.6	5.2	4.6	4.8
Yorks/Humbs	2.9	8.0	10.9	9.4	9.7
North West	8.4	16.4	14.9	14.1	14.2
East Midlands	3.5	5.0	9.5	8.4	8.5
West Midlands	7.3	8.7	10.5	10.7	10.5
South West	13.0	10.5	7.2	12.2	10.5
Eastern	18.1	14.9	11.8	12.3	12.4
Greater London	22.7	26.3	11.8	12.8	13.2
South East	21.9	5.6	18.2	15.4	16.3
Urban/rural setting					
City Centre	3.3	4.7	3.1	2.3	2.7
Urban	39.4	43.0	21.8	20.7	22.3
Suburban Residential	44.3	38.6	53.5	50.2	50.8
Rural Residential	9.3	8.7	15.2	16.6	15.7
Village Centre	2.9	4.0	3.5	3.6	3.5
Rural	0.9	0.9	2.7	6.4	4.8
Number of dwellings in vicinity					
Under 25	8.4	8.4	8.9	12.4	11.0
25-49	10.8	8.7	13.2	15.7	14.5
50-99	19.6	29.8	25.9	23.6	24.4
100-299	36.3	35.7	35.7	31.2	33.0
300-499	15.6	9.0	9.7	9.3	9.6
500+	9.0	8.4	5.5	4.9	5.3
Isolated	0.2	0.0	0.6	2.8	1.9
Base	453	323	4364	7736	12876

Source: 1991 English House Condition Survey, our analysis.

48

Table 4.12
Neighbourhood environmental conditions by equity status, owner occupiers in England 1991

| | Percentage by equity status category | | | | |
	Negative or none	Positive up to £5,000	Positive over £5,000, aged under 45	Positive over £5,000, aged 45 or more	All home owners
Abandonment/ vacancy					
No problems	79.7	82.9	87.2	90.5	88.8
Some problems	20.3	17.1	12.8	9.5	11.2
Lack of maintenance/invest ment in area					
	52.9	59.2	63.4	62.5	62.4
	47.1	40.8	36.6	37.5	37.6
Scruffy or untidy gardens					
	66.4	64.8	74.0	76.2	74.8
	33.6	35.2	26.0	23.8	25.2
Parking or traffic congestion					
	35.9	46.3	46.6	48.9	47.6
	64.1	53.7	53.4	51.1	52.4
Base	453	322	4353	7728	12856

Source: 1991 English House Condition Survey, our analysis.

of older households with positive equity seek to move (18 per cent), reflecting the tendency of this group to prefer to stay put. No clear pattern emerged from the reasons given for seeking a move, but people with negative equity were less likely to have taken specific actions such as viewing houses, or putting their own home on the market, perhaps because of the difficulties facing them in actually securing a sale. They were more likely, however, to have begun to save for the move.

Location

Table 4.11 shows the regional location of negative and low positive equity cases. Negative equity was very strongly concentrated in the South East and London, confirming the well-established picture. Three quarters of those with this problem in 1991 lived in London, the South East, the South West or East Anglia. The distribution of households with low equity formed less clear pattern. Negative equity appears to have been more of a city centre and urban phenomenon than a suburban or rural one, and one found more in larger housing neighbourhoods than in areas with only a small number of dwellings.

Table 4.12 shows variations in neighbourhood environmental conditions (as assessed by English House Condition Survey surveyors) by equity group. There is a relationship between the problems shown and equity, with those in negative or with low equity being more likely to experience problems arising from vacant properties, low levels of investment in properties and community facilities, untidy gardens and open spaces, and traffic congestion or parking nuisance.

Conclusion

The 1991 English House Condition Survey shows households in negative equity to have been, in the main, recent purchasers living in London and the south of England, with a higher than average proportion of first time buyers. Typically such households were headed by someone aged under 45, with a higher household incomes than average, often with two earners and no children. However such households were typically living in lower value and relatively older dwellings than average, but not dwellings which were significantly more likely than average to be in poor condition. The profile of those with low positive equity was generally very similar, and distinctly different from those with more equity. If older households are excluded because this group includes many retired and outright owners and many who have accumulated substantial amounts of equity through mortgage repayment and past price rises, younger households with negative equity stand out to a much lesser extent from their peers - they are mainly at an earlier stage in the lifecycle, having missed out on the opportunity to accumulate housing wealth in the heady years of the mid 1980s. In most respects those with negative equity can be characterised simply as having been in the wrong place in the housing market at the wrong time.

Notes

1   We are grateful to staff at the Department of the Environment, Transport and the Regions for the provision of previously unpublished data from the 1991 English House Condition Survey and for their assistance with the analysis and interpretation of data.
2   Where amount of equity held is based on the professional estimate.

5 Routes into negative equity

Introduction

So far, the evidence presented by us on negative equity has relied on estimates from various sources as well as data from social surveys, and has tended to focus on the extent of the problem and the characteristics of those experiencing negative equity. Inevitably, these approaches can only give a partial picture and uncover certain dimensions of the phenomenon. In order to gain a greater depth of understanding of the ways in which households got into difficulties and how they coped with these difficulties we embarked on a series of in-depth interviews in 1994 and 1996 with households from sub-samples taken from our wider research carried out in Luton and Bristol. The households were purposely selected to reflect the range of circumstances highlighted in the survey, and included those experiencing for example, high negative equity, unemployment, marital breakdown and general indebtedness. The discussion incorporates the experiences, opinions and attitudes influencing particular patterns of behaviour within a housing market which has recently been marked by boom and severe slump.

Those interviewed included both first-time buyers and previous owner occupiers. The first section of this chapter will concentrate on the motivations, circumstances and experiences of first-time buyers, both prior to and following purchase. The next section will look at the somewhat different experiences of previous owner occupiers from which a number of issues will emerge. Whilst there are similarities between these different groups, particularly in relation to general attitudes to home ownership, the housing histories of previous home owners have proved the most uneven and, in many cases, the extent and impact of negative equity has been greater. The final part of this chapter will consider particular aspects of household circumstances such as employment and financial situation, as well as relationship breakdown. The interviews highlight the increased difficulties and complications for respondents when household events coalesced with instability in the housing market.

Neither in this chapter nor the next was it thought appropriate to separate Luton and Bristol cases for discrete analysis. Our primary interest was in the variation in household circumstances and strategies rather than in the nature of Luton or Bristol as different housing or labour markets. At that level the key difference was in their initial selection as places which have experienced different degrees of price volatility as discussed earlier in the book. In that context, Luton, with much steeper price increases and falls, did in general produce more extreme cases of negative equity and sharper examples of different elements of strategies. Bristol households were more likely to fall into the 'sit and wait' group, to have lower levels of negative equity and to be more optimistic about future housing market developments.

The other point concerns the composition of the sample. Only home owners were included in the social survey. By definition, therefore, households which were in negative equity which have moved into the rental sectors or whose homes have been repossessed are not included. This means that those households with the most severe difficulties are probably excluded.

Becoming owner occupiers

Influences, attitudes and expectations

For many first time buyers developments in the housing market and the activities of financial institutions and property developers were major influences on their decisions to enter owner occupation, though not the only influence. Dissatisfaction with their current tenure, usually the private rented sector, was a common motivator combined with a desire to get on the first rung of the 'home ownership ladder'. In a number of cases the lack of alternatives, such as access to council housing, or impending homelessness due to family dispute or the termination of rental contracts, was an important push factor.

Previous tenure

Most typically first-time buyers had been private tenants. Others had lived in the parental home, tied accommodation and council housing.

Negative feelings and experiences of the private rented sector were expressed consistently by almost all respondents as a motivator for buying their own home. The constant threat of rent increases and eviction, unreliable landlords and the feeling that private renting was not a cost effective way to solve housing needs were a number a reasons given for its unpopularity.

For Ms T, "four lots of shady landlords and grotty houses" had been enough, feelings shared by Ms J:

> whenever the landlord would say he is going to come and fix something it was a real botch up job, you know, like I'll fix the wallpaper and he comes and puts sellotape down it. It was horrendous really. ... When I first moved in there I didn't have a lot of money so it was cheap, but as time went by it just got more and more run down. I wouldn't say it was a pleasurable experience; I couldn't wait to get out.

Mr DC, having rented in the past, preferred owner occupation because "you don't have to go out and ask somebody. ... And you don't have to worry about the fact that they might put up the rent ten fold next year or what have you".

An issue of particular relevance to many first-time buyers, especially in Bristol, was that the potential rental costs were little different to paying a mortgage.

Mr and Mrs C bought their older, two bedroom terraced house in 1990 and have a £38,000 mortgage:

> Basically we were renting and we realised that what we were paying in rent we could pay a mortgage. ... That was the primary reason for buying. ... We were going to agencies and things and being moved on and things like that and never knowing what was going on so we just thought it would be nice to have a home. ... We could afford somewhere on our salaries worth a lot more money but we just thought we'd get somewhere for the same because we were paying rent and it was very much a kind of 'oh that'll do' kind of move.

Mr CH had never considered renting and thought it "a trap" in the sense that rental costs inhibited the ability to accumulate a deposit and thus prevented mobility into owner occupation:

> I know people who've got caught in the trap and they have never been able to afford it [to buy]. So we thought we'd buy a house because the

money they pay for the rent is the money we pay for our mortgage. They just haven't been able to afford to save up and actually get a deposit.

Both Mr and Mrs CH had been living with their parents before buying their modern, two bedroom starter home in 1991 with a mortgage of £54,750. They had both been in secure employment and whilst their earnings had warranted a higher mortgage they had been unable to save a deposit, a problem that was overcome through the property developers incentive scheme which offered to cover the five per cent deposit. Based on the sale value of other properties in the area, they estimated that their house was currently worth between £46-48,000.

For others the move into owner occupation had been of necessity rather than choice, and the lack of availability of council housing had been a major factor. Mr and Mrs A, in their late twenties, had been living in council housing and would have been quite happy to remain in that sector if they could have obtained a transfer:

We didn't plan it [to buy] as soon as this. We thought eventually but we didn't have a choice really. We wanted to move and the council wouldn't move us so this was the only way we could get out.

Mr and Mrs F, in their early sixties, had been living in accommodation tied to employment.

We've been in school accommodation since 1978 because he was a caretaker. ... When we came to Luton we were advised to put our name on the council house list because at any time we might be out of work or if we stayed on to retirement we'd have somewhere to live. ... We put our name down, we registered again every time they kept sending us little bits of paper saying you're on a non-urgent list because you're not out of work etc. ... This went on till 1988 by which time we'd been on the council list for 12 years and we then approached the council and said ... we are urgent because my husband has got to retire through ill health and we are wanted to be out of this school house where we are.

After 12 years on the council house waiting list an inappropriate offer was made, which they refused. The couple were then informed that they would have to wait a further six months for an alternative offer, which they were unable to do. They were able to secure an eleven year mortgage and put down a five per cent deposit on a property valued at £38,500. According to Mrs F:

> The prices were going down a bit but we were on the tail end of the good days. I think the value of it now is about £34,000 so we are obviously in deep lumber.

Their difficulties were exacerbated by the fact that, because of their 11 year mortgage, the repayments were £411 per month. Mrs F, who provided the only income, was made redundant and the couple subsequently faced financial difficulties:

> We cashed in some insurances that we had and we got the mortgage right up-to-date and paid a bit in advance but we obviously are now in a situation where we are gradually creeping into arrears again.

Mr JO was another respondent who had previously lived in tied accommodation. He had worked as a caretaker and had been provided with a three bedroom house by his employer. During the seven years he lived there he installed "a new cooker ... fifteen hundred quids worth of kitchen units ... and a brand new bathroom unit". When his employer wanted to convert the house to offices he was ordered to leave and had little option but to buy, which he did in 1991 for £39,500.

> He gave me £500 to help me move and to find some place to pay the first mortgage payment and everything else, which wasn't too bad. At the time I was on my own so being a single man on what I was earning all I could get was a sort of £40,000 mortgage. It had to be a 100 per cent mortgage so it was hard to find.

Mr JO obtained a £39,500 mortgage but estimated that his property was now worth £32,000. The issue of £7,500 negative equity was compounded by the fact that his mortgage was with a company which subsequently ceased to trade and he was required to pay off the mortgage (including the negative equity) before he could move, without any prospect of carrying the debt on to another property.

For Mr C and his partner, family discord had been a major factor in their decision to become home owners. Both Mr C and his partner had been living in the parental home.

There were a few problems at home, not with my father but with his second wife; we didn't get on. And various things sort of, not so much pushed us out, but it was a case of that wasn't it?

Financial help from relatives combined with an incentive scheme from the property developers covering the initial transaction costs and half (2.5 per cent) of the five per cent deposit enabled them to purchase a one bedroom starter home in 1991 for £49,950.

Though Mr C had had no particular desire for a new property, the incentives, the convenience, the location and the issue of not "inheriting other people's problems" made the property appealing. At the time of the interview they felt sure that the value of the house was less than they paid for it, but despite this, Ms C felt "at least we're on the ladder".

Income and affordability

An obvious consideration for first-time buyers was the correlation between income and affordability. The majority of respondents purchased properties between 1989 and 1993, a period in which interest rates peaked and there was a major boom in house prices with national average prices doubling from £26,500 in 1983 to £45,850 in 1989. Thus, affordability issues and attractive incentive schemes overlapped as respondents sought to 'improve their situation' and get a foot on the ladder.

The experience of rising property values had been perceived as a real threat to entering owner occupation. Mr and Mrs MCN bought their modern, three bedroom terraced house in 1990 for £50,000 with a 100 per cent mortgage. Prices had started to drop slightly and, compared with previous years, it had seemed an opportune time to buy to avoid being squeezed out of the market again. As Mr MCN said:

Its when you're not on the ladder because before we bought it house prices were going up every week and we were panicking because we got engaged and the house prices were going up and up.

For a number of first-time buyers an obvious consideration was their ability to afford to buy. The volatile nature of house prices and interest rates meant that affordability and the current level of house prices was a finely tuned correlation and strongly influenced the choice of when and where to buy a property.

In 1989 Ms M and her ex-husband chose to buy their first property on a new housing estate in Bristol:

because at the time they were offering so many incentives for first time buyers, so I mean for this house, we got ... something like £500 cash back and I think we also got carpets throughout and help with legal fees as well. ... We wanted a new house because within the ten years if anything major went wrong we would be covered by the NHBC, because being the first house we ... didn't want something we were then going to spend lots of money on ... (Ms M).

Both Mr A and Mr D were attracted to the same estate at the end of the 1980s because of the apparent opportunities for reducing the costs of entry into home ownership through shared ownership schemes. Mr D, who moved from Yate to Bradley Stoke in 1989, felt that 'the builder was offering what at the time appeared to be a very attractive deal with shared equity which meant we could afford a place more than one rung up the property ladder' (Mr D).

Ms TH bought her two bedroom, inner city terraced house in 1990 for £45,000, with a £5,000 deposit which she had managed to save. She had felt that price constraints limited her to the bottom end of the market which left her little choice of location:

My wages were such that I could actually just about afford to buy something and I thought it was a good idea to get on the ladder because I was advised by my parents. ... I bought in this particular area because it was pretty much all I could afford. It was either a choice, at that time ... of somewhere over this side of Bristol which I didn't really know or getting a modern place in Yate [a new development in North Bristol] or something, which I didn't really want.

House prices had continued to fall and Ms K bought a similar property in May 1991 for £40,000 with a 100 per cent mortgage. As with the previous respondent, the choice of location was more or less decided by the price of property. For Ms K the feeling had been that house prices and interest rates had peaked and levelled out:

It was the right time for me to buy, you know, in terms of where my salary was and where house prices were. ... Interest rates were about 13 per cent so it was pretty high, but I felt also that it couldn't actually get any worse so if I could afford it at that stage then I could definitely afford it later on.

58

The downward trend in house prices continued and both properties at the time of the interview were thought to be worth somewhere in the region of £39,000, leaving each respondent with an estimated £1,000 of negative equity.

First foot on the ladder or road to nowhere?

As indicated in the previous section, issues of affordability combined with the desire to establish or improve themselves in the home ownership market were influential factors for households moving into owner-occupation. However, for some households a combination of events (including housing market and employment change) had not only inhibited progression in the housing market but, according to one respondent, the housing market had actually 'left them behind'. For many of the respondents it had been the timing of the purchase which had been the crucial factor. As Ms M commented on the purchase of her property on a new housing estate in Bristol in 1989

> we thought we'd bought at a reasonable time because the prices were just starting to drop ... when we came we thought we were getting a bargain, not as it turned out (Ms M).

Mr W had been looking for a property in 1989 and on visiting a show house in Bradley Stoke was told by the sales person

> I've got a clipping from the Times here and house prices are still going up. Now is the right time to buy. Having purchased a property in December 1989 he found that the housing market declined at an unbelievable rate ... no sooner were we in there, basically we found that the house was like nose-diving in price. We live in a cul-de-sac with about 12 houses around and I think about three have not been repossessed. And the rest have been repossessed (Mr W).

Mr O felt that he and his wife had 'caught it at exactly the wrong time' when they bought their one bedroom property in 1988.

> If we had waited six months ... we would have been OK. It makes me cringe to think about it ... we paid £60,000 to buy this, and at the time that was a relative bargain. And now, six months later it goes back down to £40,000. It was almost as instant as that (Mr O).

The major problem was that the couple had a two and a half year old daughter, a one bedroom house:

> £20,000 of negative equity and we are not moving very fast! It has held us back. There is no doubt about it ... I'd love to have more kids. My wife's quite happy to have more kids as well, but you know, it is not practical ... This was the first step ... A couple of years the prices will have gone up a bit and we can move on. But it hasn't worked out like that (Mr O).

Mr and Mrs C bought their property in 1988 with a £72,000 mortgage and a £3,000 deposit. 'There was like the deadline'. Along with changes in the double allowance on mortgage interest tax relief

> prices were going up and everybody was saying, if you don't buy a house now you are never going to be able to afford one because you can't save at the rate that property prices are going up, you've got to buy a house now. Gullible as we were we rushed out and bought a house ... At the time everybody was saying, buy as big a house as you can because you are going to make more money ... we thought three years and we'll move on and get a detached house or whatever ... you suddenly come down to earth with a bump and think, it doesn't work like that. We will probably stay here longer than originally planned (Mr and Mrs C).

Their property was now worth between £58,000 and £60,000. Mrs C attributed their circumstance to 'a combination of things'. Not only did they have some £12,000 of negative equity but had accumulated mortgage arrears. 'When we moved in here the mortgage repayments were something like £600 a month ... they started going up to a stage where we were paying nearly £1,000 a month mortgage payments'. Mr C had recently become a self-employed glazier, had no regular income and was severely affected by the slump in the building industry. Mrs C considered it fortunate that her salary had remained steady and relatively secure during this period. She worked in the property industry writing software packages for property companies. Originally in the estate agency side of the company 'that dropped off because people weren't selling property but then because people couldn't sell ... all these estate agents suddenly realised that if they couldn't sell people's property they could rent it out for them and manage it

Case Study I

CUMULATIVE DIFFICULTIES

Mr and Mrs Z moved from London to Luton in 1987 following their marriage. The couple would have preferred to stay in London, but house prices were going up and Luton was affordable and convenient for commuting. As first time buyers they decided to be cautious with their first property:

'I had never had a financial millstone around my neck before. I said to my husband, let's just go for the cheaper one bedroom house and he agreed. He said yes, I think that's wise and we will go on to a two bedroom later on' (Mrs Z).

The couple paid £41,000 (100 per cent endowment mortgage) for a one bedroom house and borrowed a further £1,000 to install central heating. It has recently been valued at £30,000. At the time of purchase the couple had no children and both were in secure employment.

'I was full-time nursing then and my husband [a plumber] actually had his own business to start with when we actually got married, he was doing really well. But then that was tied to the boom wasn't it and then it all started to go down with his job i.e. people didn't want bathroom suites put in. Then he had to make a change of career and actually go and do a driving course, HGV Class 1, which cost him to do it £1,000. He went on with an agency and then two years down the track got a permanent job'.

During the period between the business failing and finding a new job the couple had difficulties paying the mortgage - 'I had my income as a nurse but we found that it still wasn't enough' (Mrs Z). To compound the situation a short time after Mr Z had taken up employment as an HGV driver following retraining, he had an accident at work involving head injuries. He has recovered from the physical effects but the psychological damage was more long term.

'Because he was off work for quite some time, probably for a good six months, they laid him off, which was difficult. I was quite surprised that they put my husband off because after all it was a genuine case where he was ill. I guess lots of companies were doing that to people at that time weren't they? It was almost like a fact of life'.

It was another three months before he was able to return to agency work. The couple got into arrears, but were able to arrange a long-term solution with their building society involving converting from an endowment to a repayment mortgage and enabling them to reduce their monthly mortgage repayments and gradually reduce their arrears. With this new arrangement the couple were just able to manage. However, the building society was taken over and the couple were required to repay the £2,000 arrears immediately, which they were unable to do. They were taken to court and are currently on a suspended repossession order and paying £150 per month on top of their mortgage repayments.

'It's probably going to take a good two years to pay off the arrears. If we enjoy these interest rates it'll be ok. It's cost us a lot more money now by taking us to court, that was an extra, what, £800 I should think which will be added on to the mortgage'.

The couple now have two children, one aged five years the other almost one year old. The house has one double bedroom in which the entire family sleep. The couple would desperately like to move but are unable to do so because of £12,000 of negative equity should they sell the property, and the additional transaction costs. They have also been advised by the building society that they would need a five per cent deposit because the couple have been in arrears on their mortgage repayments. They still have outstanding arrears which would need to be cleared before the couple could move.

61

... the lettings and management side just took off'. Although Mr and Mrs C had accepted their current situation they had bitter memories of their recent experiences.

> When you sit down and think how much you've paid out in mortgage payments in eight years and how much you'd lost, its a hell of a lot of money ... I daren't sit down and think about it, it's crucifying ... You buy a house thinking you can't lose here, you can only make money. And that's what it was like at the peak of the 80s and everybody was like money grabbing and you can't lose ... it all went, what everybody said would never happen, happened. And there was nothing, absolutely nothing you could do' (Mr and Mrs C).

The changing experience of owner occupation

Nearly all those interviewed who had previously owned properties had seen them increase in value and, as a consequence, had been able to accrue various amounts of equity. Some chose to reinvest in their next property, whilst others chose other options. Their more recent experiences of home ownership have been somewhat different, as will be shown, and the extent and impact of negative equity has been very much influenced not only by changes in the housing market but by the choice of whether or not to reinvest the equity.

Mrs H had previously owned a flat with her sister which was sold in 1989 with a profit of £14,000:

> We bought the flat in the summer and by the following summer that was when everything started to go. ... We just couldn't believe how much the flat next door to us was actually going for.

A percentage of Mrs H's share of the profit had been used as a deposit for the next property which she bought with her husband for £42,750. They moved into their current property, a three bedroom Victorian terraced house, in 1990 when they thought prices had "fallen away".

> They were low and I felt they were realistically priced where a year or two before they weren't. ... Eventually you've got to grab something, you can't just say well let's hang on another year until they come down more, they might go up'.

Mr H had gone directly into home ownership after leaving his parental home, and attributes his reluctance to rent to the influences of his father.

> My dad had always encouraged me and my brother ... because he felt that property would always increase in value, which is what he experienced. My father came out of the army with very little money and he started with one house and would normally move every five or six years each time improving the house.

Although Mr and Mrs H's experiences have been somewhat different in that the value of their house has fallen to an estimated £38,000 they both agreed that they would "just as soon put ... money into property rather than rent".

Mr and Mrs R had bought their first property (a two bedroom terrace) in 1986 for £48,500, with a deposit of £4,500 accumulated from their savings and parental contributions. "We were going to get married, so you probably quite often do [buy] for that reason. Also ... house prices were shooting up then". The couple had sold the property four years later (1990) as they were relocating because of a change in employment.

> When we sold it prices had been going down again and we couldn't sell it for about a year because there were so many houses all the same all wanting exactly the same price and not really anybody wanting to buy them for quite a while. We'd dropped our original asking price about £10,000 from the original. ... I think we sold it for £58,500.

Their current property had cost £72,500. The couple had used the equity in the previous house as a deposit and had taken out a £60,000 mortgage. According to Mrs R the house would have sold for somewhere in the region of £55,000, representing a drop of £17,500, the loss of £12,000 invested from the previous house, and negative equity in the region of £5,500.

Mr U had owned two properties prior to his marriage but had sold because of changes in employment. He had not felt any particular desire to stay a home owner and had been renting a property from the Development Corporation which was satisfactory.

> What I did feel is I was really anti the government's selling the council houses and then when I was in a one bedroom house, which was owned by the Development Corporation, the same as the council, ... in 1980/81 there was this real pressure to get people to buy their council houses and I got a lot of mail and 'phone calls ... to try and persuade

me to buy my own council house. ... I bought it in the end because the mortgage was less than the rent. ... It was a 100 per cent mortgage plus, because there was discount on the house [30 to 40 per cent off market value], they added on to the mortgage the solicitor's fees and things like that. ... Not having the courage of my convictions I bought it.

The equity from the sale of the former council house had formed the deposit on the one bedroom house he bought in 1986 for around £33,000. Though Mr U sold this property in 1991 for £42,000 and therefore had a substantial amount of equity to reinvest, his choice had been to divert the capital to pay off other outstanding debts. The £6,000 deposit placed on his current property, which he had purchased in 1991 following his marriage, came from Mrs U who had been living at home and had been able to save. The couple took out a mortgage of £62,000 and estimated that the value of the property at the time of the interview was at least £6,000 less than they paid.

Though the couple had seen their deposit "disappear" they felt comparatively lucky and were aware of others in worse situations.

A friend of ours who's a solicitor, he deals virtually exclusively in house conveyancing ... is stuck in a flat on top of a carpet shop ... and he paid £44,000 for the flat and it's probably not worth £25,000. So I think we know a lot of people worse off than us. Two friends of L's, girls who bought houses at a great struggle right at the peak of the market and have now got big mortgages and houses that aren't worth the money.

Clearly the more recent experiences of previous owner occupiers had been very different from those in the past.

Mr DC had been an owner occupier in a previous marriage, but had left the relationship with no capital. Following his second marriage he had bought a property for £30,000 and sold it some 18 months later for £57,000, and moved away from the area to a bigger detached house. Because of employment relocation the couple were forced to move 12 months later and, in 1989, sold their second property for £86,000 as "the market just took off".

Mr and Mrs R moved to a new estate on the outskirts of Luton in 1988. They had sold their older property in the centre of Luton for £56,000 having paid £20,000 in 1985. 'We just poured it all into this one', and took out a £54,000 mortgage on an £80,000, three-bedroom property.' Their expectation was that in the next three years 'prices would have gone up so that we could move onto a four-bedroom detached ... on the outskirts somewhere. We were rubbing our hands with glee because it was almost like, OK, well the next house we can buy is something in the mid 100,000's ... This little voice inside your head says well, they've got to come down at some point, but you didn't expect them to come down quite the rate they did ... it just kept going down and down'.

They estimated that the property was currently worth £67,000 but felt lucky that they 'haven't actually got into negative equity like the rest of the street. We feel much healthier in that sense than some of the people around us' (Mr and Mrs R). Whilst Mr and Mrs R thought they could sell relatively easily and move on, they were concerned about transaction costs eating into the capital they have left. For Mr and Mrs R the move to a four-bedroom detached house would have to wait as they considered spending £20,000 on extending their current property 'to get what we want'.

Mr and Mrs DC purchased their current property in 1989 for £80,000. Though they had accrued equity from the previous properties, the choice was made to invest in a new car and to take a holiday. Both were in secure employment and felt they could afford the mortgage of £75,000. According to Mr DC:

> Virtually the day after we moved in the house market disappeared through the floor. We were stuck basically. ... The house is still falling away, you know. ... I would think, for a quick sale now, I'd get fifty five. ... The price of our house has actually gone down a third really.

Mr DC remained philosophical, however:

> I'm resigned to the fact that I'm going to lose twenty or whatever thousand pounds on the house. ... I made £20,000 on the house down in Chard so you know, it's swings and roundabouts. I mean, if I'd put all that in this it wouldn't have been so bad. I wouldn't have such a big mortgage ... and I could be in the position that I would sell with a small profit. But that's life, there's no going back now.

Mr and Mrs HI bought a flat in 1987, just prior to their marriage. Though neither had obtained professional qualifications at the time, their potential earning upon graduation and parental encouragement gave them the courage to purchase a property for £39,000.

My parents, they were keen. P's parents thought we were really slumming it buying a flat. I mean, goodness me, people don't buy flats! ... I suppose everybody thought that house prices would keep on going up and up and up and that if we didn't buy one then we wouldn't be able to afford one later ... I suppose once we knew that we'd have enough money to buy, that we would have two incomes and could therefore get a mortgage, ... because friends of ours, single friends at the time, were buying houses and flats together because they couldn't afford to buy them on their own.

And initially the value of the flat did increase to such an extent that Mr HI was able to pay off a £1,000 bank loan, used for the deposit, a month later by extending the mortgage.

By 1988 the flat was valued at £62,500 and both Mr and Mrs HI had qualified.

... so of course we both started earning real money, and the flat was then worth 50 per cent more than we paid for it. And we thought, right, time to move and so we put the flat on the market in September and moved in January.

They bought their present property in 1989 for £90,000 and felt they had "done quite well" having seen the price of the property reduced by £10,000 in only a month. Mrs HI estimated that the value of the property at the time of interview was in the region of £70,000. However, Mr and Mrs HI had substantial equity from their previous property and, though this had been lost, felt that the £15,000 deposit had "cushioned" them against the fall in the market. "The way I looked at it, we didn't have the money to start with so we haven't lost anything".

Negative equity and newly formed households

In some cases, members of newly formed households had brought negative equity with them from a previous property, and were also experiencing negative equity on their current property.

Mr J, a solicitor, bought his first property in March 1987 with a £52,500 endowment mortgage. Having lived there for three years, he was very unhappy with the neighbourhood and desperate to move. The house was put onto the market just as house prices were starting to drop and proved difficult to sell. Mr J took the house off the market.

I am not actively trying to sell because at the moment the house is worth probably £8,000 or £9,000 less than the mortgage. If I could get the mortgage I would sell tomorrow, but there is absolutely no way.

Instead he chose to rent the property out to tenants.

The second property was purchased in 1990 for £88,000 by Mr J and his partner with a joint mortgage. Mr J's partner had sold her house and they had been able to accumulate a £5,000 deposit.

When I phoned up the building society, it was quite funny actually, ... there was absolutely no way that our earnings could possibly justify the mortgage, because the mortgage on this house was as much basically as we could afford for it, if it had been the only one. And he said, right, it doesn't matter you've got another house even though its with them, that won't stop you renting it ... that was obviously when house prices were plummeting, they weren't bothered, they just said, oh well, we'll give you the money anyway.

The couple initially found their commitments very difficult "because the mortgage rates were high at the time and went slightly higher". With regard to renting his first property Mr J considered it "a loss-making venture because the rent doesn't really cover the mortgage. The house probably costs me about £30 a month net, which is probably better than walking away with ten grand in debt". The financial pressure had eased somewhat as "the combined mortgages are now less than one mortgage was when it was 15.4 per cent". Their present property was valued at £75,000 at the time of interview making the combined negative equity somewhere in the region of £16,000.

Mr and Mrs ME were in a similar situation. Mr ME had moved from the parental home straight into home ownership in 1987. He had lived and worked in London but was unable to afford a property there, settling for Luton which he considered more affordable. He bought a modern, two bedroom property for £55/56,000, with a £3,000 deposit and estimated that the property was worth £50,000 at the time of interview.

This amount of negative equity on one property would not have been considered a real problem for the couple as Mr ME was in secure, well paid employment. However, Mrs ME had also previously owned a property with her sister which they had bought in 1990. They estimated that they had £10,000 negative equity on that property. "We've still got the other house because we can't sell at the moment because its worth less than we paid for it ... we paid £46,000 ... and its worth about £36,000". Though her sister had "unofficially" taken over the mortgage, Mrs ME was still responsible for half the debt.

Mrs U's first move into home ownership had been to buy a brand new two bedroom house with two friends for £51,000 in 1980. At one point the value of the house reached £75,000 at which time one of the partners was bought out, taking with her a profit of £5,000. Mr and Mrs U took over the property completely in 1989, when the house was valued at £57,000, with a 100 per cent mortgage. They have recently put it on the market for £47,995 and subsequently had to reduce the price to £46,500.

One of the motivations for Mr and Mrs U to enter owner occupation had been that:

> We thought we could make money. ... My parents had always had a mortgage and we just thought it would be an investment really. Which it would have been if we'd come out with the first friend who made £5,000 then, but we didn't and it's lost money, you know, we keep losing. We've now lost £12,000.

Mrs U's feelings for the house had become very negative:

> I hate it. Because I can't do anything to it, because we wanted to do the garden or decorating, but, you think, it's not the house we want to be in, what's the point in spending the money.

Insecurity of employment and general indebtedness

For some respondents the impact of the changes in the housing market and the creation of negative equity had been accompanied by greater insecurity of employment and income. For this group the major concerns had been of an immediate nature and concerned the monthly mortgage repayments and other financial commitments.

No longer able to live with their respective parents, Mr and Mrs HO had moved into private rented accommodation. They had made repeated attempts to get council housing but without success. Disillusioned with renting because "you're having to shell money out all the time. You don't ever get anything back for it. At the end of the day they can kick you out", they had felt their only option was to buy a house.

Mr and Mrs HO moved to their current property in 1989 for £53,000 with a 100 per cent mortgage.

House prices had just started dropping off and we thought the mortgage rate was the highest it was going to go. We gambled that was the highest it was going to go and we worked it out that we could just afford that on just my salary because L wasn't working at the time. And obviously it paid off because mortgage rates came down.

Mr HO had felt his employment secure and his income sufficient to service additional unsecured credit commitments such as bank loans, store cards and catalogue accounts. Unfortunately, this increased expenditure coincided with a reduction in Mr HO's income as his overtime and basic working week were reduced. He estimated that overall his annual income had been reduced by "five grand". The couple had found it difficult to maintain their mortgage repayments.

We fell behind, two and a half payments ... they [the bank] just said we're sorry we're not paying it any more until you get your overdraft down, because I had run up about fifteen hundred pounds overdraft. ... They just stopped paying everything and they charged us for every request that got put in.

In order to generate extra income Mr HO had found part-time "cash in hand" employment and the couple had taken in a lodger. By the time of interview, Mrs HO had found relatively secure employment and their mortgage repayments had fallen from £620 a month to £350. Although their situation was not completely secure they felt that they could see "the light at the end of the tunnel".

Because of their arrears history their prospects of obtaining a mortgage in the medium term have been blighted. They have been told that they can only get a 90 per cent advance for their next property because of their arrears. Negative equity has further complicated the situation. They estimated that their property was worth £1,000 less than the value of the current mortgage.

Clearly, their concern about their negative equity was overshadowed with more general and immediate worries about income and job insecurity. These were similar concerns expressed by other respondents.

At the time of interview Mr and Mrs P were in their third property and during a seven year home ownership career have gradually seen their equity eroded. They purchased their first property in 1987 and sold with a profit of £17,000 only eight months later. In 1988 they bought their second house for an estimated £50,000 and, despite having invested a substantial proportion of the £17,000 into the property in the form of deposit and improvements, were only able to sell the property in 1991 for the original purchase price.

Their present property, which they purchased in 1991 for £60,000 with a £9,000 deposit, was estimated to be worth around £50,000.

Mrs P was in secure employment but Mr P was made redundant and although he had found alternative employment it was insecure and badly paid.

> He used to work on the snooker circuit. He's a carpenter. He used to make all snooker tables and put them up and stay on tournaments. ... He got made redundant last December. Just at Christmas! So he's just been doing bits and bobs really. He makes sheds and conservatories now. ... It's a junior job but he took it because it was just a job. It was junior money so its really low paid. ... At Christmas, he doesn't get paid when they shut down, so he's got to go temping and that, so it's a bit difficult.

Mr P's redundancy money and the reduction in interest rates had enabled them to maintain their mortgage repayments so far but Mrs P was concerned about the period when her husband had no income:

> We struggle. Every month we still haven't got any money to spare. I don't know what we're going to do in December when he doesn't get paid for two weeks. ... The major thing that worries me is M's job really. You know, trying to get a decent job again. That's the worrying thing really for the new year. ... It's just jobs more than anything else.

Mr A had moved from a council house following a break-up with his partner. Having established a new relationship and with no chance of obtaining council housing he and his partner had wanted to 'do a little better' and bought a property in 1991. They entered into a shared purchase scheme with the builder who covered 15 per cent of the purchase price, to be repaid after two years, whilst Mr and Mrs A took out a £40,750, 100 per cent mortgage on the remaining 75 per cent. The arrangement was based on the assumption that house prices would rise and income increase. However, soon after purchasing the property Mr A, a carpenter who had benefited from the boom of the 1980s, was made redundant and work and income had been scarce and uneven ever since.

> When I was made redundant I was actually physically signed on about six different agencies. There were a few times I was doing 36 hour stretches because the work was there, another time I may not have any

work for seven days, ten days ... The wife, she had a full-time job. I mean we were getting nothing, no other help from unemployment or nobody.

In addition, they estimated the property was now worth in the region of £41,000, equivalent to their mortgage, but leaving them with an outstanding unsecured charge to be repaid to the builders. During this difficult financial period Mr and Mrs A had been required to repay the builders and had missed a number of mortgage repayments. Mr A recalled that the builders 'were quite amenable to start with and then they said we want our money and that's it. We had to go out on a limb and said right ... you can't have your money because of the situation at the moment'. Mr and Mrs A were taken to the Crown Court.

It would have been as much as £11,000. We managed to get it down to £6,000. But I mean it involves the court costs and everything. They really wanted blood. We have got to pay them so much a month and that's going to be like my lifetime and anybody else who wants to follow ... never pay that off.

The mortgage arrears precipitated an additional court case at a time when Mr and Mrs A 'were literally days away from losing the house, I mean days away from it, and the judge stood up and ... gave us a right telling off saying, this must be paid, this must be paid monthly at the correct date. And losing your home in this day ad age ... you think where are you going to go? I can't raise enough money to move into another property, even a housing association property, because you are looking at £1,200 maybe plus removal costs'.

Both Mr and Mrs A were now in full-time employment and were meeting the monthly repayments. They estimated that it would 'take three years to clear what we owe. Payments will then drop to a reasonable level. There is no way we are going to get the money back on the house. We would like to move out ... It's just bad luck for us this house, but we can't, we just got to stick at it and hope for the best'.

The amplification of difficulty

The interviews with home-owners revealed the wide range of circumstances and problems which households had experienced. For most, a variety of events coalesced at a particular time to create a particular set of difficulties.

Case Study II

> **SINGLE AND IMMOBILE**
>
> Since the early 1980s Ms F has experienced a series of job-related moves. Indeed, her move to Luton in 1989 was because of company relocation. At the time Ms F had a property in Brighton which she had bought in 1986. She put the property on the market when prices 'had stopped moving up... and it was very difficult to sell. It was on the market for about twelve months or so but I did get the money back on it'.
>
> She bought her one bedroom property in Luton in 1990, after having rented a property nearby on the estate for nearly eighteen months. The property was all that Mrs F could afford and she liked the estate because it was convenient. 'I didn't like the idea of living in Luton town centre. It's a bit ropy ... some of the areas are not too safe. Mrs F paid £48,000 with a 100 per cent mortgage and estimates that the property is now worth between £30-35,000.
>
> 'It's a shame that the negative equity's there because I do feel tied and apart from having the house here I haven't got any ties to Luton and I've lived all over. I've always been fairly mobile, I've gone where the work has been or where the inclination takes me ... It's been the first time that I have felt completely stuck with a property and also work as well. Because the job market flattened because so many people were being made redundant I felt I couldn't move jobs. If I wanted to go away, I couldn't sell so I just seemed hemmed in. I suppose I'm the sort of person who doesn't particularly like that. I like to be able to move around when I feel like it'.
>
> These feelings affected Ms F's attitude towards her home. 'I just hated it. I went through a phase where I couldn't believe I'd bought it'. Ms F was confronted with a situation in which she could 'either try and move and it would cost me £10,000 or I could so something to the property which would cost me less but would make it better for me'. She decided to invest £5,000 in the property and build a conservatory. Whist increasing her own enjoyment of the property she recognised that this may not add value to the property. However, she thinks 'it might at least make it a little bit more saleable if somebody was looking for a one bedroom. It does have the edge'.
>
> The property next to Ms F's has been rented for the last twelve months, a situation which has been problematic mainly because of noise problems. 'I've been to Noise Nuisance about it. She's a nurse, he's a DJ - it's like a combination from hell! She works odd shifts as well so she can be coming in at three in the morning and the music goes on ... However, it's not just the music that is the problem but the construction of the property and, in particular, the staircase. 'The walls are very thin; and the staircase, because it's metal, its seems to act almost like an aerial - if you're walking up the stairs the whole building shakes'.
>
> Ms F has accepted that, for the time being, she will remain in the Luton area. She has been looking for employment opportunities in the area and hopes to find something more suitable and secure in the near future. She has also been contemplating strategies for improving her current housing situation. The owners of the property next door are keen to sell as soon as prices rise a little. Since Ms F's salary has increased she has had the idea 'that perhaps if they would sell it cheap enough then I could buy it and knock it through and then I've got a three-bedroom house and no neighbours! It would make a decent size three-bedroom house with a nice big garden' and would also broaden the potential market when she comes to sell.

In a period of stable or rising house prices options for coping with difficult life events may be varied. In a period of uncertainty these difficulties become intensified and options may be more restricted. This was particularly apparent amongst households experiencing relationship breakdown. Financial and other difficulties are not unusual following relationship breakdown. However, for a number of respondents the high initial financial commitment required to purchase a property during the period of rapidly rising house prices, followed by the slump had compounded and complicated the situation.

Ms M and her ex-husband bought their property in 1989 for £55,000 with a five per cent deposit.

> At the time I was a trainee solicitor and so my income was fairly low and my husband didn't have a huge income. I mean this was when building societies were throwing money at you ... (Ms M).

In 1991 Ms M's marriage broke up.

> I'd been made redundant and my husband left me when my daughter was just three weeks old, so I had no job, so I was on income support and then we had awful problems with the building society because they wouldn't let him take his name off the mortgage to begin with because I was on benefits. And then when I did go back to work, I had to be at work for six months before they'd transfer it and all the arrears that had built up, ... about £500, had to be paid off ... I didn't get any money back from him or anything like that over it, so he walked straight into a new relationship, buying a new house with no negative equity and left me with this over my head, you know with no immediate prospect of getting rid of it (Ms M).

She continued

> I think most of my problems were caused by the fact that our marriage broke up, because realistically, if I'd still been with my husband in a stable relationship we would have been able to got out of it, because of my income and his, we should have been able to get out of the negative equity trap fairly quickly I would have thought.

Ms M was now in secure employment and her sole income has quadrupled. Although not the ideal solution, Ms M intended taking out a negative equity mortgage.

I'll still be carrying negative equity because I'll have taken out a loan of 120 per cent on a house and likewise, if I try to save up enough money just to clear the negative equity, I would then still be starting off in my mid 30's as a first time buyer. So either way I don't win but at least I get to move onto a bigger house which was always the plan.

Mr D bought his property in Bradley Stoke through a shared ownership scheme with builders on a 75/25 basis. The arrangement for the 25 per cent 'was that any interest on it was scaled over a period - it was interest free for a while and then there was a token interest for a year, slightly more, until year five the full amount became payable' (Mr D). At the end of this period, Mr D owed the builders £17,500 (the builders wrote off £3,000 interest) whilst the value of the house had fallen by £16,000-17,000. In addition, Mr D's £52,000 mortgage was taken out on the basis of two incomes.

This house I bought with my ex-girlfriend. It may have just been bad timing but the relationship didn't work out very shortly after buying this place so my problem was compounded by the fact that the property market dipped dramatically but also that what had been bought on a joint mortgage I was financing myself ... After I split up, there were three or four months of horrible, money going all over the place and not knowing what's going on. I think at one stage the arrears were around £2,500 when the mortgage was around £500 per month so it was five months worth of arrears. Then fortunately interest rates started to tumble and I agreed with the building society that I'd continue my repayments at £500 per month even when the mortgage by then was just under £400. I paid back about £1,000 or £1,100 of the arrears but there was still a big chunk of arrears, £1,500. This amount was added to his mortgage 'which originally stood at £52,500 and now stands at £54,000 or thereabouts' (Mr D).

So, in addition to the arrears Mr D had a substantial amount of unsecured debt. 'She cut and run; I can't make her pay me back the negative equity - I'll take it all on myself. I'm stuck with the problem' (Mr D).

Conclusion

The diversity of circumstances of the respondents reflects the heterogeneity of people in owner occupation and the uneven and diverse impact of housing market change. One of the overriding features, however, was a pervasive belief in home ownership as part of 'tradition' and 'culture', and a lack of confidence in the private rented sector. However, increasing caution, feelings of distrust, relationship problems all point to the psychological and emotional dimensions experienced by households during this period of difficulty.

Whilst in housing terms the prospects of most respondents have diminished, they have experienced increased incomes. For another group there was a sense of a general decline in prosperity, with negative equity accompanied by lower, or less secure income resulting from changes in the labour market. There was also a cohort of people who felt they had been left behind and had failed to meet their housing aspirations.

What is clear is that the extent of the problem and the nature of solutions cannot be assumed simply by concentrating on the level of negative equity. Those with the largest negative equity are, in the main, those with relatively secure and rising incomes. Whilst lower negative equity may appear less of an issue, it was most likely to accompany less secure and uneven earnings. To gain a meaningful picture it is vital to view negative equity within the context in which it occurs. It is not simply the extent of negative equity which is important but the circumstances of the household and their ability to overcome it.

6 Coping strategies among home owners with negative equity

The previous chapter drew on a series of in-depth interviews with households in Luton and Bristol to explore their housing histories and current circumstances. Having established the diversity of household routes into the situation and circumstances in which they now find themselves, this chapter describes the strategies adopted to cope with negative equity. It looks first at the coping strategies adopted by households in negative equity in 1994. It then goes on to revisit these strategies drawing on a series of interviews carried out in 1997. The chapter then goes on to discuss whether attitudes towards home ownership appear to have changed in the light of recent experiences.

It is appropriate to make some general observations before focusing on specific features of the different households in the study. First, the term 'coping strategy' indicates a degree of conscious planning in adverse circumstances. Inevitably, some strategies are more conscious than others. Some households have a clear, premeditated housing 'plan' which is adjusted as external or household circumstances change. For others the strategy is a series of ad hoc responses to difficulties with no evident goal other than to maintain their current housing status. Housing is, however, a central component in peoples' lives and for most households the maintenance of housing status and security is a key priority. Coping strategies need to be articulated with the idea of a housing career. We tend to work implicitly (and often explicitly) with a model of the home ownership market which implies a series of moves as part of an upward trajectory or associated with life-cycle changes. Coping strategies, however, are specific and often short-term adaptations to contingencies with little reference to longer-term plans. The implication in the present context is that negative equity is requiring many households to make such short-term adaptations and such measures may be distorting normal housing career trajectories.

Second, it should already be evident that aggregate statistics relating to estimates of numbers of households in negative equity conceal enormous variations in the nature and severity of the problems involved. For some households a very modest shortfall between current property value and outstanding debt can represent a major problem: for example, if jobs are insecure, household income is low, savings are limited and the property is becoming overcrowded. Conversely, substantial negative equity may be an inconvenience rather than an overriding difficulty in a household with rising and secure income, substantial savings and with no real pressure to move. For some households the strategies adopted involve a high degree of choice whereas others may be characterised mainly by constraint with very limited room for manoeuvre.

Third, various strategies are outlined below. For some households there is no real strategy associated with negative equity. There may be the usual lifetime aspirations of moving up market into a bigger house in a more desirable neighbourhood but negative equity is not seen as a particular complication or something which needs to be addressed in the short term. For some households the existence of negative equity is merely an additional problem within an array of difficulties and in that sense the strategy is a general one rather than consciously targeted on the loan/value gap. Other responses to negative equity include renting out the property, moving to rented accommodation, trading down, paying back into positive equity, investing in and extending the current dwelling, or simply waiting till prices begin to rise. Much depends on the push factors involved which mainly relate to job situation and family growth. The most common response has been to sit tight and wait in the hope that price inflation would lift the property value back into a positive equity situation. The longer it takes for this to happen the greater will be the number of households faced with the need for job mobility or which are trapped in dwellings they consider inappropriate for their needs. It should also be emphasised that although the various strategies are discussed separately, most typically the households interviewed had considered or pursued a combination of more than one.

Buying into positive equity

Many of the households interviewed had experienced rising real incomes and falling housing costs as mortgage interest rates had been reduced. In many cases households were in a position to service a substantially larger mortgage but were locked into their present dwelling because of negative equity. Mr and Mrs C, for example, were second-time buyers in Bristol who simply wanted to move back to the countryside. They were slightly unusual since there was no necessity for a move for either job, family growth or employment reasons. They had explored the various schemes that had been introduced by lenders to aid mobility for households in negative equity. This quickly proved a dead end:

> When the government announced that they were relaxing the guidelines for people with negative equity on mortgages I went round Bristol and trawled through the mortgage lenders and nobody was interested. One of them actually had a scheme and they said they had managed to help one client and transfer the mortgage to a new house but they said it took six months. So basically no one's interested. They've got the money, they've got the interest coming in, so they're not interested in the people who are suffering.

They had considered selling up and renting while they paid back the debt but felt that rental costs were so high for a reasonable family property that they be would worse off than having a mortgage. Their preferred strategy had been to save and pay back lump sums at regular intervals. An original mortgage of £75,000 had now been reduced to £64,000 through lump sum repayments "but the house is still falling away, you know … I would think I'd get fifty five for a quick sale". And he went on:

> I'm trying to get it down. I mean, I kept sort of paying money off and then going to the estate agents and saying what's the value now and every time I pay off five thousand pounds the house drops five thousand, so I wasn't getting anywhere. But it's been relatively stable for the last 12 months so I think I'm actually winning slowly.

Despite this experience Mr and Mrs C remained committed to home ownership and indeed to taking out a large mortgage on their next house. They were both in secure, public sector jobs (he was a police constable) and saw that as relieving them of any real anxieties about mortgage repayments.

Mr and Mrs R in Luton had adopted a similar approach when all others had failed and had paid back a large sum as well as remortgaging. This was a more complicated story and involved periods of real difficulties in making mortgage repayments, loss of a second income through child-birth and change of job. Again they were second-time buyers who had taken out a relatively large mortgage on their next house. This was a new detached house on an estate. They had paid £98,000, borrowed £87,000, and it had been valued recently for £70,000. They had tried to sell but there was little interest so they had decided to reduce their mortgage costs through remortgaging. This was more complicated than they had anticipated. They had a fixed rate mortgage at 12 per cent and had paid a £3,000 arrangement fee to enable them to change the mortgage at a later date without penalty.

> What we didn't bank on was the house falling below the value of the mortgage. Once we got into a negative equity situation which was about the same time as interest rates fell below 12 per cent and we wanted to change we couldn't, because we didn't have the money to pay off the negative equity. It's only in the last couple of months that we have managed to save up - well my husband has. He started up his own company and managed to take some money out of it. He also did a private contract and got paid quite a large lump sum for it so we now have enough money to pay of the negative equity on the mortgage.

This way of escaping from the problem was, however, seen as painful and a waste of money. Their strategy was to pay off what they could in a lump sum (around £15,000), remortgage at the normal variable rate and take out a bank loan to cover the outstanding debt. Their monthly mortgage payments would fall from £800 to £250 and they could then easily service a £5,000 loan from the bank. They were bitter about the valuation of their property at £70,000 by the building society when a local agent had recently valued it at £8,000 more. But they were most disenchanted at the loss of a substantial sum of money for no apparent gain.

> At the moment we've got this big lump sum - we've never been in a position where we've had so much cash. My husband is quite depressed about the fact that it is just going to go down a black hole basically. Our mortgage is going to be a lot less so we are going to see a benefit but it doesn't feel like it.

The feeling that paying off some of the mortgage to reduce the negative equity was money down the drain was also expressed by a couple, both accountants, who had considered this option. They were considering a job move to Scotland and he would qualify for a lump sum through a voluntary severance scheme. They had contemplated using the severance money and savings to pay off the debt.

We were going to pay some of that over into the mortgage ... so that when we came to sell it wouldn't be such a huge burden. But it's a bit galling in that when the mortgage rate was 15 per cent we were breaking our backs to pay it and now we are saving up money to pay off more. We could have been saving for a nice holiday. It's a bit depressing when you think of how much money you have shelled out.

This approach to the difficulties of negative equity was associated with previous owners rather than first-time buyers and to those with reasonably high household incomes who could save large amounts or had the opportunity to acquire lump sums through their employment. It was also a strategy associated with substantial negative equity where drastic measures were required. To sit and wait for house prices to lift the value of the property out of negative equity was perceived as a lengthy and uncertain process. However, the households adopting this strategy were under no pressure to move for job or family reasons.

Delayed and constrained mobility

Has negative equity had an impact on patterns of mobility? There are examples among our households of people who need to move and are unable to do so. Constrained mobility is, however, less common than a more general expression of delayed progression up the housing ladder. Younger childless couples, or young single people with, say, £1,000-3,000 of negative equity were often apparently unconcerned about the shortfall but felt that under different circumstances they would have moved on earlier. Typically, they were in situations where rising real incomes and falling housing costs meant they could service a larger loan. One young couple in Bristol were clear that they would have moved if property prices had risen rather than fallen.

We're both in the situation where we've got a lot better jobs now than when we moved here, we are both earning more money and could afford a bigger mortgage but we are stuck with this because we don't want to lose on it. So we're just really sitting for a while until, you know, you get quite complacent really don't you. Let it run, let's see its course, let's see what the economy is going to do and then maybe we'll jump back into the housing market.

This attitude and this situation were representative of the younger first-time buyer group in Bristol. Compared with Luton, negative equity was generally less severe and houses had been more affordable. It is this group of reasonably well resourced, often double earner households at an early stage in the family life-cycle which will escape from negative equity when there is a modest revival in the housing market.

Within the case study group there were also those who were having to move for job reasons and where employers were intervening to ease the difficulties of negative equity. In both cases they were professional households which qualified for relocation packages. This involved a guaranteed sale at a price determined by the average of three valuations. However, employers were not prepared to cover the whole of the shortfall between the sale price and any outstanding debt. For two young accountants, Mr and Mrs K, this involved a £10,000 debt and they decided not to move, although not simply because of negative equity. And the company was also having problems developing policies for relocation in a depressed housing market.

> By the time they came round and did the valuations and everything I was getting fed up with work anyway. The thing is, with the restructuring, the business hadn't really developed its relocation policy; to me it was kind of making it up as it went along because of course it was encountering all sorts of problems like this one because of course when you go along and say to them 'I'm going to make a £10,000 loss on my house - are you going to give me the money?', they sort of think 'crikey! £10,000 for what'? And because it was a compulsory transfer they couldn't tie you in to stay with the company for two years or anything like that. So they were thinking if she goes next year we've lost it. And lots of people didn't want to make the move anyway because even if they didn't have negative equity they were making a big loss on the sale which the company wasn't going to reimburse them for.

A similar scheme was on offer to two scientific officers in the civil service who were being relocated to the north of England from Luton. Again there was a guaranteed sale and sale price. Any negative equity, which in their case was £15-20,000 was covered by an interest free loan.

Trading down

One way of moving was to trade down to a smaller and/or cheaper dwelling. For one household this was part of a recouping strategy, a way of making good the loss on the current dwelling. For another it was the only way out of an increasingly difficult situation.

Mr and Mrs H were expecting their second child and were in a small two bedroom house on a new estate in Luton. They had contemplated building an extension but the garden was small and they had no savings. With negative equity they had no borrowing capacity so the only way to acquire more space was to move house. Mrs H also disliked the area and the local schools. They had tried to sell two years earlier in 1991 but there was little interest. Since then prices had fallen further and their negative equity was now around £12,000. They had put their house back on the market at £48,000 in February 1993. In December, at the time of the interview it was still unsold and they had reduced the price to £46,000. The outstanding mortgage was £57,000. Some prospective buyers were now emerging and they were hopeful of a sale in the near future.

Assuming they could sell their house the only option for them in the owner occupied sector (other than staying put in what they considered to be an extremely undesirable situation) was to buy a cheaper house in a cheaper area. They had only one modest income coming in which limited their capacity to save and to finance a larger mortgage. Their building society, the Halifax, were offering a scheme which enabled them to carry the unsecured debt with them, albeit at a slightly higher rate of interest.

> We got some details of houses ... If you bought one for say £40,000 and put our £12,000 on top we'd have like a £52,000 mortgage for a £40,000 house. As you go further north they're cheaper, so we're looking where the cheaper houses are.

And because they needed three bedrooms they were looking for the kinds of houses which offered best value for money.

> We've got to look at cheap things like former council houses, things like that, which isn't really what I want, but you know, to get out of here really.

For Mrs H the strategy they were adopting was one of despair and disappointment. She saw their situation of negative equity stretching well into the future and fundamentally affecting their future housing market trajectory.

> So every house we have will have negative equity. It will take years and years. I mean our dream house we can never have really, because we've got too much negative equity.

For Mr LB, a builder with £7,000 of negative equity, trading down in price terms was a way of acquiring a larger dwelling in need of refurbishment on which he expected to recoup his losses. 'Sweat equity' would replace the negative equity. His strategy was to buy an unmodernised house for about £30,000 which he expected to be able to sell after improvement and repair for about £55,000. The financing arrangement was rather complicated but he regarded himself as a skilled housing market entrepreneur.

> Well, I've spoken to the bank and I can get a personal loan, not a bridging loan, it will be a personal loan for a fixed period so you actually borrow the money for two years and you have to do the modernisation in that two year period, rather than a bridging loan which is open ended. Then you can get it on a fixed rate of interest, so it's not too bad. Then the loan is obviously secured against the mortgage and once the house is worth the money it should be, you can then get a revaluation and your mortgage will be extended. It's a bit messy but I can't see any other way of getting out of it really unless you are just going to sit here and wait another three or four years until they [prices] have picked up again.

He also had firm views on the state of the housing market as being paralysed by institutional intransigence and the inability of people to see things in relative terms He thought that if there was a general shift in lending behaviour to enable people to carry negative equity with them, the housing market would start moving and the problem would disappear quickly. And for him, money gains through the housing market were something of a mirage.

> I never have been able to get into that kind of thing. You're not making money really, relatively speaking you're standing still. Because everything is moving at the same rate unless you're going to sell at a ridiculous price and then move down a scale. The relative values are meaningless. It's very difficult to explain that to people,

they just don't want to see it. They think they've lost money and really they haven't because the house they want is less than it was going to be.

The rental strategy

Renting in various forms was seen by many as a way of coping with the current difficulties in home ownership. The idea of renting out the current dwelling was particularly evident in Bristol where a number of the younger single people and couples in cheaper, inner city properties had anticipated this as a way of coping with the current inflexibility in the home ownership market. Ms V expected to be working abroad for a while and intended to rent out the house. She thought that property prices would rise in her absence and lift her out of negative equity. Another respondent, Ms T, wanted to live with her boyfriend in Swindon. He also owned a house with negative equity so she would rent her property until housing market circumstances changed.

There was also a view that rents had continued to increase while house prices and mortgage costs had fallen. Therefore, the rental income from a small terraced house was likely to easily cover the mortgage payments and any maintenance costs. There was also a recognition that the greater difficulties faced by some young people in gaining access to home ownership because of stricter eligibility rules and lower percentage of value loans - and a greater degree of caution about purchasing in present circumstances - had boosted demand for renting. Obviously, inner city properties with a ready rental market among students and other young people were better located in this respect than newly built dwellings on the edge of Luton or Bristol. Even here, however, there was some indication that repossessions and arrears cases were contributing to a growing demand for family dwellings to rent. In other words, there was some evidence that the difficulties faced by some households in gaining (or regaining) access to home ownership were creating the solutions for others to escape situations of negative equity.

There are a variety of problems associated with renting out a property to enable mobility in situations of negative equity. Most fundamentally, unless the owner is also going to rent accommodation to live in, there are difficulties raising a loan legally for another dwelling, no tax relief on the mortgage interest and complications with capital gains when the property is resold. Purchasing another dwelling is only an option for those with incomes sufficient to service a much larger loan. The accountants discussed in the previous chapter were in this category. They had rented out their first purchase and traded up to a larger dwelling with a second mortgage. One of the consequences was an increase in negative equity since prices had continued to fall after the second purchase.

Others were deterred by the management fees and a negative view of tenants.

> By the time you add on your commission fees to the estate agent you'd be very lucky to get your money back and then if you get a family that trashes the place suddenly you've lost everything.

Housing association activity in some of the areas in the study was also providing an escape route. The only elderly household included in the study were hoping that a local association would purchase their former council flat which had proved extremely difficult to sell. They were also contemplating a permanent return to renting, probably from a housing association, and a move from Luton to the south coast to be nearer relatives. The negative equity on their property was small, around £2,000, but this was a major problem with their limited pension income. Unless house prices rose they expected to have to wait for another two years for an insurance policy to mature. They could then sell at a loss and pay back the difference.

Housing association purchase on a new build estate on the edge of Luton was enabling some households to move while creating various social tensions. In a situation where an upward movement in house prices could be critical for residential mobility and removal of debt, the conversion of some properties from owner occupied to renting was perceived as a very negative influence. One person's solution could, it was believed, exacerbate the difficulties of negative equity for someone else. As Mr B of Luton remarked:

> If I sell this to a housing association, which really reads council house, it must affect the possibility of [others] selling if they need to.

He continued

> It becomes almost like a council estate. Not that I'm against council estates but if you are going to spend £60,000 on a house you don't really want someone living next door to you who possibly doesn't give a bugger what the place looks like.

Investing in the current dwelling

A variation on staying put and waiting is to invest more positively in the current dwelling rather than moving. If the dwelling is large enough for the household's needs or has the potential to be enlarged, this is an option. Among households in small starter homes or in flats, the sit and wait strategy was generally associated with investing the minimum. And there was a widespread and understandable view that to spend much on the fabric of the dwelling in a situation of negative equity was likely to involve throwing good money after bad.

> We did think at one point of putting in a fitted kitchen and things like that didn't we, and double glazing and all that, but we just thought well we are making a loss on the house anyway so what is the point (Mr and Mrs OK).

> I mean if I'd the money I would have put French windows in the back. It just seemed pointless putting more money into the house when it was losing money all the time. I expect we would have done if the equity had gone up (Ms T).

Some dwellings and gardens were in any case so small that there was no scope for extensions. Others could have created more space but had no borrowing capacity and no savings.

> I mean if we wanted a further advance to say get our bathroom done or get our windows done, we can't because we have not got enough equity in the property to be able to do it (Mr and Mrs M).

One household was cashing in some British Telecom shares to build a substantial extension which would add three rooms. The impact of negative equity on improvement and repairs is, therefore, somewhat ambiguous. In some circumstances, building an extension serves as an alternative to moving to a larger dwelling. Households may have surplus funds because of rising incomes and reduced mortgage costs. On balance, however, the impact of negative equity is more likely to have reduced investment in the basic fabric of dwellings. Much of that activity occurs on acquiring or selling a property and negative equity has contributed to reduced mobility. And the propensity to carry out more than minimum repairs must be less in a situation where property values are falling, many households have limited borrowing capacity and any returns on such investments are uncertain.

Revisiting coping strategies

The previous section identified a range of coping strategies adopted by households in negative equity in 1994. The commonest strategy at that time was to 'sit and wait' in the hope that house price inflation would lift the property value back into positive equity. There was still an optimism amongst householders that this was a temporary aberration and that soon the housing market would be back on course. Though this was still the preferred strategy for some respondents in 1997, the combination of declining optimism and the inability to delay moving has resulted in a range of more complex strategies being adopted.

Mr and Mrs X bought their property in Bristol in 1991, and after having a child decided the house was too small and they needed to move. They put the house on the market for 12 months but were unable to sell, even at a price that was some £12,000 less than they had originally paid. They had investigated various ways of moving with negative equity but found most of the schemes to be too restricted and inappropriate for their particular circumstances, as they only applied where the negative equity was around five per cent of the total value of the property. In their case it was substantially more. They became aware of a scheme which enabled them to buy a property which had to be over £60,000, and requiring a 15 per cent deposit. Mrs X had recently taken £16,000 voluntary severance from her job in an insurance company which provided the deposit. They intended pursuing this option whilst renting out their current property, the rent from which would just about cover the mortgage. They had also converted from an endowment to a repayment mortgage in the hope that any rise in house prices would eventually reduce the negative equity enabling them to sell the property.

Mr and Mrs W were keen to sell their Bristol property as they wanted children but were concerned about whether they could 'afford a £400 per month mortgage with my wife not working and with a child'. The couple had approximately £10,000 of negative equity on the house which for Mr W meant that he was

> paying a mortgage all the time and I'm not making any money on the place. I mean it is no nearer being mine in five years time than it is now ...'. The couple took the decision that 'now might be the time. We've got to try and get out of it somehow - out of this situation ... so what we've done is we've bought a house through a shared ownership property, through my wife, over in South Wales - she's got a half a share in it. ... She is never going to have to buy the other share if she

doesn't want to and we can live there indefinitely. Our intentions are eventually, when we can do something with the place in Bradley Stoke, we'll take the other half of the house on (Mr W).

Case Study III

COMPLICATED EXIT

In 1989 Mr W and his former wife bought a two-bedroom, mid-terrace property in Bradley Stoke for £54,000 with a 100 per cent mortgage. The mortgage was calculated on the basis of three and a half times their joint incomes and, as an incentive, the builders gave the couple a discount on their mortgage for the first three years, gradually reducing the amount until the fourth year when the full mortgage payment came into affect.

Six months after purchasing the property the couple 'found that the house was nose-diving in price. The house is in a cul-de-sac with about 12 houses around and I think about three have not been repossessed ... the rest have'. The property is now valued at £42,000. Twelve months after purchasing the property the couple split up. By this time the monthly mortgage repayments were 'up around £500-£550. She moved out and didn't want anything to do with the place and said "just hand in the keys ...".' Mr W, now with sole responsibility, had considerable difficulties paying the mortgage.

'I was working in a supermarket because of course the housing market crashed and so did my work. I'm a plumber you see. So I ended up working in a supermarket earning £400-£500 per month. My mortgage is £500 per month. Obviously I've got to eat, I've got a car, I've got to pay the bills. What can I do? Mr W decided that his only option was to move out and rent the property. 'So I got what I could afford which was basically a bedsit. It was almost as bad as squatting. It had damp running off the walls. Permanently cold all the time'.

Whilst living in rented accommodation Mr W formed another relationship. When the couple finally considered living together Mr W felt his rented bedsit to be unsuitable and the only affordable option seemed to be to move back into his original property. By this time the mortgage was £320 per month, comparable to the cost of decent rental accommodation.

The couple decided they would like to have children but felt that the two-bedroom property was not big enough and were also concerned about their ability to maintain the mortgage payments 'with my wife not working and with a child'. Even though Mr W had returned to work as a plumber following an improvement in the building industry he felt cautious about the extent of the recovery. 'The building industry has been a bit up and down. You only want a month out of work and you'd be in trouble'.

The couple decided 'we've got to try and get out of it somehow ... so what we've done is we've bought a house through a shared ownership property through my wife in South Wales. She's got a half-share in it. She is never going to have to buy the other share if she doesn't want to and we can live there indefinitely ... Our intention eventually, when we can do something with the place in Bradley Stoke, is to take the other share of the house on'. The property was valued at £40,000. For their half share the couple paid £20,000 with a mortgage repayment of £90 per month. They pay £110 rental. Their total monthly payments amount to £200 which they feel they can afford quite comfortably. The property in Bradley Stoke, which they are unable to sell because of the £12,000 negative equity is rented out for £380 per month which just covers the cost of the mortgage.

Mr and Mrs Z, who lived in a one-bedroom property in Luton with their two children also saw shared ownership as a potential way out of their housing difficulties - mortgage arrears and £12,000 of negative equity. They had investigated purchasing a two bedroom property for £20,000 from a Housing Association through a shared ownership scheme, which would have left them with a mortgage of £35,000 absorbing the negative equity and the arrears. Unfortunately the transaction fell through and properties with this type of arrangement 'are not as plentiful as you would like (Mrs Z). Nevertheless, the Z's were encouraged as 'we could have actually afforded to move because ... that would have paid off the arrears, it would have helped us out of a scrape'. Mrs Z continues to 'listen out' for suitable opportunities.

Tenure diversification, one of the major changes on the case study estates in Bristol and Luton, can be seen not only as a product of the changing patterns and location of employment (the high-tech industries of Aztec West, the relocation of the MOD, university expansion) but also be seen as a response by home-owners to problems of negative equity or more general housing problems. Mr D rented out rooms in his Bristol house to generate income. 'In my experience, the people that rent have been relocated because of their jobs - that's a big one - with Aztec West and so on around here. I've had a couple of students who were at UWE [University of the West of England]' (Mr D). For those seeking employment and educational opportunities, or prohibited through repossession from purchasing, rented accommodation on the estates provides an immediate, relatively good quality solution. For existing home owners wanting to overcome housing difficulties renting out their property has become a potential way out.

> Luton is a University town now see, [and] they've taken over ... what was a factory behind Sainsbury's ... to do with car maintenance courses and things like that. There's one opposite the paper shop ..., a four-bedroom, has got four or five students' (Mr D). Mrs D continues 'there's also people that have lost properties earlier on in the slump and can't afford or cannot get mortgages so they're just moving around, because the short term lets have been on offer at the moment' (Mr and Mrs S).

For Mr and Mrs HO it was the rental strategy which offered a potential solution to their housing difficulties. 'You pay another per cent of your mortgage for doing it and you obviously pay management fees to whoever rents it out for you. We wouldn't make any money by doing that but it would get us to the area that we would want to be in and hopefully if ever prices did come up again maybe we would break even or ... [put] a bit towards our negative equity (Mr and Mrs HO).

The study indicated a growing awareness of negative equity mortgages amongst households, but also revealed feelings of dissatisfaction with the various schemes on offer. In general they were felt to be inflexible and inappropriate.

Mr and Mrs HO bought their current property for £73,000 and have a mortgage of £71,000. They estimated the property was worth in the region of £60,000-£63,000. Mr and Mrs HO wanted to move and felt they could afford to take on a larger mortgage, but they are unable to because of the difficulty 'for us finding £12,000. We have put it on the market before and we sort of enquired about borrowing enough to cover the negative equity'. Referring to negative equity schemes Mr HO commented

> there is a catch in all of them. You have got to have so much savings, you've got to pay a slightly higher interest rate because you're borrowing more than the property is worth. So, they are not really giving you anything for nothing ... there are cons with it ... (Mr HO).

He reflected on how the schemes did not really fit his needs and expressed increased caution at taking on this additional, long-term financial commitment in an unpredictable housing market.

> ... we could afford a £100,000 mortgage but we could only look for a property for £88,000 because £12,000 of it is lost money. ... We couldn't get what we wanted for £88,000. Well, not as nice as we wanted and we just, you know, didn't want to borrow that money because then obviously if there was another drop in the house market already our house that we were in would be in negative equity ... so then we just thought well we will just sort of sit here and try and save a bit towards it so at least hopefully next year if we can move we'll have a bit behind us this time (Mr and Mrs HO).

Ms M, a single parent with a good income, consulted her bank regarding a negative equity scheme in which 'they will lend you up to 120 per cent of the value of the new house, you have to save up five per cent deposit'. However,

> the stumbling block is that my daughter is in full time nursery at the moment and they won't consider it whilst I've got £500 a month going out in nursery fees. So until she goes to school next September we're

still stuck. I mean having said that, I haven't saved up the deposit yet...

She expressed dissatisfaction with the lack of help and advice she had received in relation to the negative equity.

I've had to do it myself and have constantly come up against brick walls apart from the last six months when I had ... a quite hefty salary increase and suddenly everybody that slammed doors in my face when I asked them first time around, I wouldn't say bending over backwards, but they are trying to help. Which annoys me in a way ... it was just a case of what's your income, well you don't stand a chance. Not what can we do in the future ... (Ms M).

Mr and Mrs O lived with their two and a half year old daughter in a one bedroom flat, a situation which put 'severe family pressure on the three of us' (Mr O). They had investigated a variety of schemes but had found none of them helpful. 'We looked at part exchanging this place and buying another place, or even renting somewhere else. But basically the Building Society are not keen on the negative equity situation'. Mr O, previously a trainee chartered accountant, has been made redundant twice during the last eight and a half years, whilst Mrs O was made redundant from the National Health Service. Because of reduced income the couple build up some arrears. According to Mr O 'We are with Nationwide and if you have any history of arrears at all - regardless of the reasons - they won't give you a chance'. Mr O was aware that there might have been a slight upturn in the market but felt it would make little difference to his situation. 'It may be one per cent - but when you are looking at £20,000 your are looking for 50 per cent increase in the value of your house before it will start'.

The O's current strategy had been to change from an endowment to a repayment mortgage in order to 'accelerate the payments to try and wipe out some of the negative equity - but that's obviously going to take time as well'. Mr O was now in secure employment and Mrs O hoped to return to the labour market on completion of her undergraduate degree. 'We have set ourselves up a four year plan. By the end of the four year period - in four years time we will have paid off the negative equity and be in a position to move'.

Educated Home Owners

The complexity of strategies mentioned above indicated a growing awareness of available options, and a confidence and determination amongst households when entering into financial transactions. However, negative experiences of home ownership had also contributed to an increasing level of caution. Ms M, who intended taking out a negative equity mortgage in the near future, had been far from satisfied with the advice received from her building society and financial adviser.

Case Study IV

MAKING THE MOST OF ADVERSITY

Mr and Mrs C bought their second property, a three-bedroom house, in 1988 for £75,000 with a £3,000 deposit. Their previous property had been a one-bedroom house in the next street bought a year earlier for £55,000. They were able to sell this property for £58,000 which meant they 'just about broke even'. Their current property, on which they had a mortgage of £72,000 is now valued at £58,000. The couple were given a lump sum of £8,000 by Mr C's parents which enabled them to reduce their negative equity to £6,000. At the same time they switched to a fixed rate mortgage for five years.

The couple bought their current property because they wanted 'a bigger house. Salaries went up and at that time we could afford a bigger mortgage. Until the interest rates went through the roof. At one time we were paying £1,000 per month' (Mrs C). At the same time Mr C, a self-employed glazier and suffering from the slump in the building industry, could no longer rely on a regular income. For Mrs C, however, the fluctuations in the housing market were to create opportunities which were crucial in ensuring continuity of employment and thus enabling the couple to maintain their mortgage repayments:

'We write software packages for property companies. I was doing sales of an estate agency product at the worst point of the market and it was terrible, I hated it. You know, cold calling, trying to sell estate agent software when they are not selling anything. But then all these estate agents suddenly realised that if they couldn't sell people's property they could rent it out for them and manage it. So all these companies that were doing estate agency started doing lettings and property management. So that which was a smaller part of the business, the lettings and management side, just took off, so I moved to that side of the business. It was quite lucky really because if the company had just been in estate agency we would have gone out of business'.

Though obviously unhappy about their current situation Mrs C is particularly thankful 'for the fact that we'd moved from a one-bedroom house because I saw a lot of people in one-bedroom houses that they were desperate to get out of and they couldn't get out. If we'd have stayed in that one-bedroom house that would have dropped in price and we would have never sold it and we would have killed each other I think if we'd have been still in a one-bedroom house. So I can at least say, well at least we are in a house and we are happy here and we don't want to move ...' It's a three-bedroom house 'there is only the two of us and the animals, so it [the negative equity] isn't really a problem at the moment'.

92

Now I think, sod the lot of them. I shall do it myself and figure out what's best for me when the time comes and look at it in a great more detail. ... I think when I go into my next house purchase, I'll be looking to find somewhere that I can stay for a long, long time in case it happens again (Ms M).

Mr and Mrs R felt that they

are definitely more aware. I think it is reassuring that people are going through the same thing, because there's always the danger that you can make a silly mistake and fall foul ... But some people study things in a lot more detail and understand you know, this is the best type of mortgage for this circumstance, and it is quite interesting listening to them and what their choice has been and why they've chosen that, you know, you store it up. Yes, I would use people's experience and knowledge to my advantage (Mr and Mrs R).

Mr and Mr A, as first-time buyers felt they were

at the mercy of the wolves. There are so many, not loopholes, but so many ways they can get you that they don't tell you about. I don't know if there is anybody you can go to for help ... I mean you're talking big money. It's the biggest outlay, financial outlay, of your life isn't it? It's a young town, getting bigger by the day. There is a lot of movement, there is houses going up all the time. Now they know more about it and because of the building crash, those people that are moving in have, hopefully, got a better chance of getting on (Mr A).

Ms F, however, was not so sure about the prospects for future purchasers

I suppose that like a lot of people I don't ever see the market forces being exactly as they were during the 80s. I would think that people could get caught because the new buyers behind me that didn't go through the spiral ... eventually you're going to have managers in mortgage companies and building societies who didn't go through the spiral so I can see quite easily that they would actually ... on bonuses and everything else ... I won't be one of them I don't think! I'll have a few more grey hairs then, I'll have a little bit more sense' (Ms F).

The comfort of strangers and friends

Apocryphal tales or not, households with negative equity can usually find someone worse off than themselves, or at least others in similar situations. This is an important element of the coping strategies employed. In the housing boom there was always someone who had made a killing and moved rapidly up market. Now there is usually an even greater casualty of the housing market recession. They may be brothers, sisters or 'the couple down the street'.

> So many of my friends have got houses that are worth less ... there's somebody I know in Montpelier [Bristol] who bought a house for something like £70,000 and now it's worth £48,000 or something and it's just ridiculous. I think there's so many people, especially my age, in the same boat, so you sort of shrug your shoulders and think, well there are so many people who have been affected that you don't get worked up about it really (Ms H, Bristol).

> We've got a couple of friends in Portsmouth and they paid £57,000 for a three bedroom terrace. It's literally eleven foot wide. In fact the size of the house is probably the size of this room. And that's on the market for what, £46,000 (Mr and Mrs McC, Bristol).

> I know there's loads of people in the same situation as us if not worse, people in studio flats who've got a baby or whatever and that's worse. I mean, at least we've got two bedrooms and a house, because a flat, I imagine, would be a lot worse (Mrs H, Bristol).

> But it's not just us it's happened to, there's plenty people up the street with the same problem ... there are a few people up the street that paid £80,000 for a house like this and it's only worth £52,000, so I mean they really are in trouble (Mr and Mrs M, Luton).

> I mean the houses sold down the road for £61,000 and they were £85,000, so they've lost an absolute fortune ... and the people down that street where the houses were worth £125,000 and now they are going for £78,000 so everybody has got the same.

And the same respondent commented on the drastic actions that some of these less fortunate people had apparently taken:

People actually just disappear overnight ... he did on the corner there. Christmas Day he moved out. And the chap down the road straight after Christmas. They both moved, left these houses and got new houses, new mortgages before they were repossessed (Mr N, Luton).

No doubt there are strategies available which involve deception and fraud and which may lead to prosecution or credit blacklisting at some later date.

Changing attitudes?

Had the experience of negative equity changed attitudes towards home ownership? In general, preferences for home ownership and expectations of remaining in that tenure remained apparently unaffected by recent experiences. There were views expressed about the lack of alternatives on offer or the unattractiveness of these alternatives, and there were apparently changed expectations about the nature of home ownership. In other words, home owners appeared less inclined to stress the investment and accumulative aspects of home ownership and to stress the use value of the dwelling. And those with an experience of private renting saw home ownership as an escape from a tenure which lacked security and control and which they tended to associate with inferior quality, difficulties with landlords and wasted money.

Attitudes towards private renting were, however, contradictory. Some households felt that they had moved into home ownership prematurely and a later entry would have been more sensible in terms of their financial and family circumstances. Moreover, some respondents who were very negative about their own experiences of private renting were themselves contemplating renting their current dwellings to others as a way of coping with negative equity.

There were also views expressed by younger households about the relative security of council housing which was available to parents and others in the previous generation. As Mr and Mrs Y observe,

Its like your Mum and Dad, they say 'we had hard times, we were on orange crates when we moved in'. But they had a council house. They had perfect security there even if Dad lost his job, right, the house would always be paid for. They always had somewhere to live but ... if either of us loses our jobs we haven't got a house to live in.

It was also striking how many parents of younger buyers had bought their council houses at a discount and were seen as privileged. One respondent was in a dilapidated Victorian terrace in the middle of Luton with a few thousand pounds of negative equity. His brother was trapped in a studio flat with a wife and a young baby and substantial negative equity. Meanwhile his mother was living alone in a three bedroom, semi-detached council house which she had bought with maximum discount when she was given a lump sum on retirement. And his grandmother was living in a similar dwelling which she had bought from the council and which had been the family home.

Mr W, who felt bitter about his recent experiences of marital breakup, repossession and negative equity (in his new home), saw home ownership as necessary for social status but council housing as the privileged point of entry.

> You are definitely a lot better off I think with a council house - financially you're quids in aren't you? If my parents had had a council house I probably would have gone into one. I do know it works like that, if their parents had a council house they end up having one because obviously they realise it's a better deal. Especially when you can buy it for half price. That is making me more bitter than the mortgage trap.

One of the strong impressions from talking to younger owners who had entered the housing market in a period of recession and uncertainty, and who were in situations of negative equity, was that they were very aware of their changed housing circumstances when compared with parental experiences. Parents either had long histories as home owners and had amassed substantial positive equity or had grown up in an expanding council sector with subsequent opportunities to buy well built dwellings at less than market value. Our younger buyers had no expectations (nor in most cases an expressed desire) of gaining access to council housing. Equally, they did not see their parents' experience of home ownership as indicative of their own.

Whether or not attitudes towards the investment aspects of home ownership had changed was difficult to determine. There was certainly a much greater emphasis on the house as home rather than as something to be bought and sold in the market when conditions were most propitious. The general message from the in-depth interviews was that respondents still aspired to owning their own home but that these aspirations were no longer so closely associated with financial gain.

When we first bought it was the old 'invest in property' line, you know you can never go wrong if you had money in bricks and mortar. And yes sure, it paid off for us because we sold in the heights ... and bought in the heights unfortunately, but the money we've lost on this place is to us just paper money, not real money for us at all. And we're not in negative equity. But I would not see a property as an investment quite as much now (Mr and Mrs R).

The notion of developing a housing career, of accumulating equity and improving one's dwelling, also seemed to have diminished with households expecting to make long-term housing choices involving manageable financial commitment.

There was very much that culture if you get married you buy a house and if you can't afford to buy a house then you should wait until you can. There has been many a time before I got divorced, when we were quite happily married, that we wondered what the hell we were doing with a house and wished we had stayed renting and just put the money aside. But at the end of the day you're stuck with it, you just have to make the most of a bad thing really. ... When I go into my next house purchase ... I'll be looking to find somewhere that I can stay for a long, long time in case it happens again. And I will also go into it thinking about what would happen if interest rates go up and things like that which is not something that really we thought about when we first bought ... I would want to know that I could cope with it on my own if it happened again, you know, if the interest rates went up (Ms M).

Mr and Mrs HO felt their recent experiences had left them feeling:

... Once bitten twice shy I think is the trouble with us know. We are just really scared to make that move again because we don't think that buying has done us any favours at all. We have lost £21,000, £9,000 that we put into this and the £12,000 that we now owe (Mr and Mrs HO).

The feeling that the severity of the fluctuations in the housing market could have been avoided was a theme articulated by many respondents. Following his experiences Mr A thought that 'the government has got a lot to do with it and I still believe that they could have seen that yes, the prices were going up and up and up and they were going crazy. I don't know a lot about politics but surely they could have said, whoa. They control the mortgage rates, they do control that, so surely they could have said, you know, maximum increase, instead of the bottom just going wallop like, you know. There must be people throughout the country, not just in isolated areas, who are suffering' (Mr A).

However, these attitudes could change if market conditions change. A new house price boom could produce the same frenzied excitement. Equally, commentators in the late 1980s were quite content to extrapolate into the future from a very specific set of market conditions and to assume that attitudes at that point were set in concrete, or at least in appreciating bricks and mortar. For example, in Saunders' (1990) analysis of the investment orientation of home owners in the late 1980s, he found strong evidence of a "deliberate and coherent investment strategy through the housing market" and uses various quotes to demonstrate the point. One respondent is quoted: "People have a housing strategy in the back of their minds. We've been here 15 years and people say to me 'You've missed a move, you should have traded up years ago'" (p.199).

Given the circumstances of the households in this study it would be wrong to claim that their views were representative of all home owners in the early 1990s. Nevertheless, there were strong contrasts between the discourse around home ownership then compared to a few years previously. Again Saunders (1990) provided his own neat description of the situation in the late 1980s: "Go into most pubs in most English towns on most nights of the week and you will likely overhear a conversation about how much people now believe their housing is worth and whether now is the time to trade up or cash in" (p.202). People are, perhaps, more likely to swap stories of success rather than perceived failure. The fact that many of the households in this study emphasised use rather than investment aspects of housing could well be a post hoc rationalisation of the situation in which found themselves. It does not necessarily justify a view that people are less likely to be concerned with the investment aspects of home ownership and that a significant shift of attitudes has occurred. Nevertheless the quotes below contrast with the views of the households quoted in Saunders' (1990) study.

If we'd had a choice we would like to have rented and saved rather than having had to borrow, because we had to borrow money straight away from the start to get this anyway because we didn't have anything. No, if we'd had a choice we'd have rented for a year or two. A bit more like they do on the continent, there's no pressure there to buy when you're in your twenties, people don't expect to buy until they're in their thirties. Ideally it would have been a council house. I mean a lot of our friends and relations are very mercenary and they all went into it [home ownership] thinking they were going to make a quick buck. We're not like that, we want somewhere to call home (Mrs G).

We just want this house and then to move to another in a few years and stay there so we've got a firm base and it's home. I don't really see it as an investment, a way to make money (Mr and Mrs CH).

I just wanted my own place. I just wanted to live on my own and I wanted to do all that I wanted. I like painting and I like DIY stuff and I just wanted my own place. I don't really think there was much of a financial thing behind it all really.

The same respondent goes on to contrast her views with those of some of her friends:

They are not frantic money makers, I think that it's just happened you know. It's like so many other people, they were looking to move and then they realised that the house was worth quite a lot of money, I don't think you make hard cash out of sales if you go to another house. I think you just set your sights even higher each time with what you can afford (Ms T).

I mean we might have different perceptions but I just look at a house as somewhere to live and I never thought like other people think that it's a way of making money. I think when we looked at this place we didn't think 'oh we'll make some money on it', we just thought we'd like to live here, we can afford it, that was the main thing (Mr and Mrs U).

Someone in negative equity may be disinclined to admit that they purchased believing that they were going to make a lot of money. Those kinds of attitudes do seem to be associated with 'other people'. What emerges is, however, an awareness of the home ownership market as being something of a lottery. You do well if you're in the right place at the right time. It is an accident rather than the product of a conscious investment strategy. The investment aspects appear to be less central to people's motivation than in, say, Saunders' (1990) account. There is, however, a strong belief that a mortgage is buying something in a way that rental payments are not. Even in negative equity there is an ownership goal, a debt is hopefully being reduced and at the end of the day there will be something to show for it.

7 Changing attitudes and behaviour

Introduction

As negative equity continued well into the 1990s as a problem facing many home owners, policymakers and analysts became concerned about the longer term impacts on attitudes and housing market behaviour. To what extent was a longstanding recession, a discourse of risk and uncertainty and a widespread lived experience of unsecured debt breeding more deep seated disillusionment with home ownership. These issues have been touched upon in earlier chapters drawing on qualitative research on a small group of households living in new build estates in Luton and Bristol (Forrest, Kennett and Leather, 1997). The general message from this research was that even those worst affected by negative equity still aspired to home ownership as the preferred option. These aspirations were, however, no longer so closely associated with financial gain. The notion of developing a housing career and of accumulating equity also seemed to have diminished with households expecting to make longer term choices involving manageable financial commitments. This cohort of households which had purchased their dwellings in the late 1980s and early 1990s felt they had been left behind and had failed to fulfil what they had come to expect as normal housing aspirations. Increasing caution, feelings of mistrust and relationship stress were all evident in this group.

This evidence on the attitudes and behaviour of home owners most affected by negative equity was limited, however, in two ways. First it was based on a small number of qualitative interviews. Second, it referred almost exclusively to home owners living in new estates in two locations, Bristol and Luton. Alternatively, more general evidence on housing attitudes tended to be either out of date (Department of the Environment, 1994) or referred to home owners as whole (for example, Jowell et al., 1997). What was required was a more representative cross section of the cohort of owners which entered the market as prices began to dip. The English House Condition Survey (EHCS) was the best source for such a sample. The 1991 EHCS was the sixth in the series and involved an initial sample of around 25,000 dwellings and combined a physical and social survey. As indicated in Chapter 4, for the purposes of this study the important advantage of the EHCS over other sources is that the calculations to determine levels of housing equity are based on professional valuations rather than self assessment or crude estimates from general house price trends. Essentially, our social survey which was carried out in early 1997 was a random sample of households identified by the 1991 EHCS as having low or negative equity. From 284 issues addresses nationally, 157 full interviews were achieved. On key variables such as household type and age, the characteristics of the households included in the survey were very similar to those of the group described in Chapter 4. In other words, the survey data are representative of the kinds of households identified as having low or negative equity from the 1991 EHCS.

Changing behaviour

Among households with low or negative equity it appeared that mobility, investment in the dwelling, employment and household formation had all been variously affected. The impacts were most striking in relation to the negative or nil equity group. For the sample as a whole respondents were likely to state that the housing market situation had encouraged them to spend more rather than less on their home, but the opposite was the case for those with negative or low equity. And perhaps most strikingly, 14 per cent of this group stated that they had postponed having more children because of the uncertainties surrounding the market value of their home. The policy implications from these kinds of data are rather ambiguous. For example, it may be that restricted mobility encouraged a net increase in investment in repairs and improvements to dwellings.

Although half of the respondents with negative equity stated that the housing market had not impacted on them in any of ways listed in Table 7.1 a substantial minority had clearly been affected. Although the sample number was relatively small, if these percentages were applied to all households in negative or low equity situations it would amount to a significant figure.

Table 7.1

Current housing equity[1] by market value of home has had various impacts[2]

Sample	Negative 73	Low[3] 58	Moderate (24)	All 157
	%	%	%	%
Had property on market and withdrew it	8	9	8	8
Postponed move	27	9	4	17
Discouraged looking for job/turned down job offer	8	3	-	5
Affected decision to have more children	14	2	4	8
Spent more on home	12	22	21	17
Spent less on home	16	5	-	10
None of these	53	60	79	61

Note[1]: Owner's estimates
Note[2]: Multicode possible therefore totals do not add to 100
Note[3]: Low equity = up to £5,000; moderate = £5-10,000

How had this group coped with the uncertainties of the housing market? Consistent with our previous research the most common response was that they had been simply waiting for better times - for house prices to rise (Table 7.2). It was also evident that these home owners had been more inclined to shop around for different kinds of mortgages in a changed market and to actively and consciously pay a higher monthly payment than necessary or save to increase their housing equity. Some one in 10 of all households in the sample stated they had been saving in order to reduce their outstanding loan and some had cashed in endowment or savings plans for that purpose. These numbers are more striking if Holmans and Frosztega's (1996) estimates are used to provide a national figure. They could suggest, for example, that some 120,000 households who were or who had been in negative equity had been saving to reduce their outstanding loan.

Table 7.2
Coping strategies in a changed housing market

Sample	Negative 73	Low 58	Moderate (24)	All 157
	%	%	%	%
Paying more off mortgage than required by lender	5	5	4	6
Waiting for prices to rise	40	21	4	27
Saving to reduce outstanding loan	14	12	-	11
Changed type of mortgage	7	14	13	10
Extended/converted home to provide more space	3	12	8	8
Cashed in endowment policy or other savings plan to reduce mortgage	3	3	-	3
None of these	48	45	71	50

Note: Multicode possible

House price expectations

Most of the sample had been fairly optimistic about future movements in house prices when they had moved into their current dwelling. As Table 7.3 shows, one in 10 felt prices were still falling, around 60 per cent had seen prices as static, and a quarter had felt prices were rising. Those with negative equity were more likely to have envisaged future price rises, and hence presumably to have been willing to commit themselves heavily in terms of mortgage payments and less cautious about purchase price. Surprisingly, households in negative equity were also slightly more likely to have anticipated future price falls.

103

Table 7.3
Expectations of future house price changes when moved in

Expected future house prices changes when moved in	Negative	Low	Moderate	All
	%	%	%	%
Had risen, would go higher	29	21	21	24
Had risen, would not rise further	25	21	21	22
Had fallen, would not fall further	32	41	33	36
Had fallen, would fall further	11	9	8	10
None of these/don't know	4	9	16	8

As regards the changing state of the residential property market in 1997 (Table 7.4), three quarters of respondents felt that the market generally had picked up, although significantly, less than two thirds thought that it had improved in **their** area, perhaps indicating the extent to which views at that time had been influenced by press reports of market recovery rather than personal experience. However the nature of the recovery was seen more in terms of increased turnover than price rises. Two thirds thought more properties were selling in their area, but only 10 per cent thought prices were going up a lot and less than half thought they were rising a little. Those with negative equity were neither more nor less optimistic than average. In this context some relevant comparisons can be made with data from the 1994/5 Survey of English Housing. In that survey, 28 per cent of owners thought house prices would remain stable over the next five years and 75 per cent expected prices to rise. Only 3 per cent expected further falls.

Looking into the future, respondents were more optimistic about future price changes with over a quarter thinking house prices in the next five years would rise by more than the rate of inflation. Those with negative equity, and with low positive equity, were least optimistic, as were frustrated movers and those whose household circumstances had changed. Older respondents aged 50 and over were particularly optimistic, perhaps reflecting the fact that they had seen prices recover after previous down-turns, whereas those in their 20s had not.

Table 7.4
Expectations about the housing market

Percentage of households in row category	Resp under 30	Resp over 50	No full time worker	Neg equity	Low pos equity	Frust mover	Econ loss	House-hold change	Diffs with housg costs	Lost money on house	All respon-dents
Base	56	10	11	73	58	74	54	65	53	19	157
House price movements over next 5 years											
Rise more than inflation	18	60	27	23	26	20	26	22	30	5	27
Rise same as inflation	32	0	9	29	26	32	26	35	28	26	29
Rise less than inflation	25	10	18	25	29	22	30	20	21	32	24
Stay the same	13	20	45	19	10	19	13	17	17	37	13
Fall	5	0	0	3	2	3	4	2	0	0	2
Don't know	7	10	0	1	7	4	2	5	4	0	6
The market											
Market generally has improved recently	71	90	45	73	69	62	69	68	66	53	73
The market in this area has improved recently	63	80	45	62	57	57	57	63	53	47	62
More properties are selling in this area now	57	80	64	68	52	61	65	65	60	53	64
House prices are going up a lot	7	0	18	10	3	11	11	11	13	5	9

Frustrated movers

A group of 74 households were characterised as 'frustrated movers'. Of these, 21 (out of 157) had unsuccessfully tried to move and a further 53 had not tried but said that they would have liked to move. Table 7.5 shows some characteristics of these households and their responses to various attitudinal questions compared to other households in the sample.

These frustrated movers were more likely to be in the 30-44 age group, and to be couples with one or two children or larger households than to be younger single people or childless couples. They were likely to have been resident since before 1990, and less likely to have two or more full-time earners in the household. However, in terms of the income of the chief income earner, they were not significantly different from other households. They were more likely to be first time buyers, to have experienced some form of domestic change such as the birth of children or a relationship problem, and to have experienced some form of financial difficulty in terms of meeting housing costs, or some form of change in their economic circumstances associated with unemployment, redundancy, or business failure. They were much more likely to be in negative equity and to regard this as a problem.

These problems had had a clear impact on attitudes amongst the frustrated movers. They were much more likely to be dissatisfied with their home, less likely to regard the home as good value for money, and more broadly, were much less likely to be pleased to be a home owner. They were also more likely to feel that owning was too much of a responsibility, and less likely to feel that property was a good investment. They were also more pessimistic about recent changes in the housing market, either generally or in their local area. Being a frustrated mover did not, however, appear to lead households to take a negative attitude towards investment in repair and improvement. They were more likely than others to have invested in their dwelling, either in order to increase the chances of sale, or to help to deal with the changing requirements which had led them to wish to move in the first place.

The impact of the housing market recession and negative equity on the capacity to move can thus be regarded as one of the main problems for home owners and one which has done most to reduce positive views about home ownership. It has particularly impacted on younger households with children, either because family formation or expansion have led to stronger pressures to move to obtain more space or a more suitable dwelling (with a garden for children to play in for example), or because such households have been less able to afford a move as a result of reduced earning capacity.

Table 7.5
Frustrated movers: characteristics and attitudes

	Frustrated movers %	Others %	All respondents %
Age of respondent			
Under 30	30	41	36
30-44	64	48	55
45 and over	8	5	7
Household type			
Single under 65	18	18	18
Couple under 65	19	41	31
Small family/single parent	43	28	35
Large family/large adult	20	13	17
Income of CIE			
Under £200pw	13	14	14
£200-499pw	73	69	71
£500pw and over	14	17	15
Number of full time earners			
None	10	5	7
One	50	45	47
Two or more	40	50	46
When moved in			
Pre 1988	10	5	7
1988-90	42	15	27
Post 1990	48	70	66
First time buyer	69	54	61
Social change	47	22	34
Financial difficulty	60	42	50
Economic change	45	25	34
Negative equity problem	24	7	15
Investor in dwelling	60	43	51
Dissatisfied with home	14	2	8
Home good value for money	55	86	71
Pleased to be home owner	84	96	90
Housing market has improved recently	62	82	73
Housing market has improved in this area	57	66	62
Owning is too much of a responsibility	16	5	10
Property is a good investment	73	82	78
Base	74	83	157

Victims of redundancy have also suffered because negative equity has reduced their capacity to relocate to find a new job.

Views on the home

The survey found a routine high level of satisfaction with the home with less than 1 in 10 respondents expressing any dissatisfaction. For England as a whole, some 6 per cent of owners with a mortgage expressed some level of dissatisfaction with their home, (ONS, 1996). Dissatisfaction levels were not therefore significantly above the national average. Households with no earners, and those with negative or nil equity were the most likely to be dissatisfied (almost one in five), with frustrated movers, young respondents, households which had experienced changed circumstances, and those experiencing difficulties with housing costs also more likely than average to be dissatisfied. Our previous research (Forrest, Kennett and Leather, 1994; 1997) had also shown that negative equity affected satisfaction with the home.

Around three quarters of respondents were pleased that they had bought their home, with older respondents and those with low positive equity most likely to take this view. A similar overall proportion thought that buying their home had been good value for money. Those most likely to disagree with this were households with negative equity, those with no full time earner, frustrated movers, and those who had had difficulties in meeting their housing costs. Even amongst those who said that they had lost money on their home, more than a half still said that buying had been good value for money, presumably when set against the alternatives. Overall, less than 20 per cent felt that they had lost money on their home, and less than one in ten felt that their home would be more difficult to sell than others in the area. Those without a full time earner, those in negative equity, and those who had experienced difficulties with housing costs were the most likely to feel that they had lost money.

Attitudes towards home ownership

Despite the long recession in the property market most people felt that the attractiveness of home ownership had not changed. Indeed , one in five felt it had increased (Table 7.6). Those most likely to feel that ownership had become less attractive were those who had lost money or experienced economic problems, especially those without an earner in the household.

The vast majority, especially older home owners, were pleased that they had purchased a dwelling. This group are perhaps better placed to take a long run view of the undulations in property values. Again predictably, those without a full time earner in the household, those who reported difficulty in meeting housing costs, those in negative equity, and frustrated movers were least positive.

Attitudes were, however, less positive towards housing as an investment. Only just over half felt that housing was now as good an investment as ever. Those who had experienced problems with housing costs, frustrated movers, and those with negative equity were the most likely to feel that housing had become a less attractive investment. Moreover, a third of respondents said that they were now more likely to save or invest a windfall than to use it to trade up into a more expensive house. A similar proportion said that they were more likely to stay put even if they could afford to move and more likely to spend surplus income than to trade up. The greatest enthusiasm for alternative uses for spare funds came from older respondents, who were perhaps more likely to have such funds and less likely to need to trade up, and from those who felt they had lost money on their existing house.

Over a number of years the British Social Attitudes Survey has asked people how they would advise a hypothetical newly-married couple, both in secure jobs, about house purchase. The same question was asked of our group. Nearly 75 per cent took the view that they should buy as soon as possible. Most enthusiastic were older people, those with negative equity, and those who had experienced changes in household composition. A quarter of respondents would advise the couple to wait and then buy but hardly anyone advised the couple not to buy at all. Households with no full time earner were particularly prominent amongst those advising the couple to rent.

Respondents were then asked to examine a range of statements about owning and renting (Table 7.7) drawn from the Housing Attitudes Survey (HAS) (Department of the Environment, 1994), and also used in the study of households in negative equity in new peripheral estates in Luton and Bristol.

The main points which emerge from Table 7.7 are:

- respondents in this study and those in Bristol/Luton were more likely to feel that owning was a risk for people without secure jobs than those responding in 1993 to the HAS. This is perhaps not surprising given the extended recession in the housing market over the post-1993 period and the increasing public awareness of the issue of job insecurity;

- some three quarters of respondents felt that property was a good investment, fewer than HAS in 1993 but more than in the Bristol/Luton study in 1996, perhaps reflecting views on the recovery of the housing market;

- only about a third of respondents felt that there were not enough homes for private renting, a much lower proportion than found by the HAS;

- around three quarters of respondents felt that that repairs and maintenance were very expensive for home owners, and two thirds felt that looking after the home was time-consuming, higher proportions than either HAS, or more significantly, the Bristol/Luton study which had only included new dwellings.

Open ended questions in the social survey delved further into the perceived advantages of owning and renting and the extent to which views had changed after a prolonged recession and the widespread and well publicised phenomenon of negative equity. Responses to these questions indicated that there had not been a dramatic change in attitudes to home ownership. It remained strongly associated with security, investment and social status and, in general, was considered preferable to renting. These findings were not particularly surprising. What was more striking was that virtually all respondents, whatever their recent housing experiences, expressed concern about the greater responsibility, cost and unpredictability involved with owning your own home. This was related not only to changes in interest rates and property prices, but insecurity in the labour market and ability to maintain financial commitments.

Asked to consider the advantages of buying their own home (see Box 1) respondents often referred to aspects of control over living space in terms of decoration, keeping pets, privacy and use of space. Buying a home was still felt to be a 'reasonable investment', and was considered to offer more choice than the rented sector which according to one respondent 'is much smaller with fewer properties to choose from'. Owning your own home was clearly connected to social status and was thought to be 'a step on the ladder'.

Whilst a small number of respondents indicated that there were no disadvantages to buying your own home, for most respondents it was the responsibility and risk of home ownership which were considered to be disadvantageous in what was more generally perceived to be a more volatile and less predictable housing market (see Box 2). As one respondent (with more than £5,000 of positive equity) commented 'if you borrow a lot of money it's a big commitment. If you lose your job or prices go down it can

Table 7.6
Attitudes towards home ownership

Percentage of households in row category	Resp under 30	Resp over 50	No full time worker	Neg equity	Low pos equity	Frust mover	Econ loss	House-hold change	Diffs with housg costs	Lost money on house	All respondents
Base	56	10	11	73	58	74	54	65	53	19	157
Attractiveness of home ownership											
More attractive	23	20	0	15	29	15	13	20	21	16	21
Has not changed	63	80	45	62	60	68	63	60	53	68	62
Less attractive	11	0	55	22	10	18	24	18	26	16	15
Pleased became an owner occupier	91	100	73	84	97	84	91	91	81	89	90
Housing as an investment											
Housing as good an investment as ever	66	80	36	49	55	45	52	48	40	47	55
More likely now to invest a windfall than trade up	30	60	36	36	34	34	31	29	42	53	34
More likely now to stay than move even if could afford to	25	60	36	30	38	27	30	26	32	47	35
More likely now to spend surplus income than trade up	36	40	55	38	41	36	35	37	40	58	41
Advise newly married couple to buy ASAP	70	80	55	70	67	65	70	72	60	53	70

111

Table 7.7
Attitudes to owning and renting

Percentage of households in row category	Resp under 30	Resp over 50	No full time worker	Neg equity	Low pos equity	Frust mover	Diffs with housg costs	Lost money on house	All respon-dents	England 1993 (SEH)	Bristol/ Luton 1996
Base	56	10	11	73	58	74	53	19	157	2194	526
Over time, buying is cheaper than renting	93	80	82	82	93	84	85	79	87	86	83
Owning is a risk for people without secure jobs	89	70	82	89	95	89	91	95	90	84	93
Repairs and maintenance are very expensive for home owners	77	80	91	75	72	73	79	89	72	67	48
There are not enough homes for private renting	30	40	55	37	45	39	43	37	39	67	27
Owning a home is too much of a responsibility	9	10	45	12	9	16	13	11	10	9	9
The only way you can get a nice place is to be an owner occupier	23	40	9	19	24	22	26	26	23	42	34
Owner-occupiers have to spend a lot of time looking after their property	75	60	64	64	67	65	74	68	65	65	58
Property is a good investment	86	90	45	73	81	73	74	79	78	92	58
More people would rent if better quality rented housing were available	45	50	55	52	48	49	55	47	49	53	46

be a great burden'. The costs of home ownership were also considered to be a disadvantage - 'it does cost a lot more than you think'. There were references to unexpected bills (new roofs, subsidence), and one respondent was 'afraid of mortgage payments'. The problems of mobility in the home ownership sector were also referred to by a number of respondents reflecting again the difficulties of buying and selling which have been more prevalent in the recent past.

Box 1

What do you consider to be the advantages of buying your own home?
Illustrative verbatim statements

I've always believed in investment in bricks and mortar
If you improve it over the years and the market price goes up you can make money
Its an investment
You can rent it out
It's your own
It will be paid off
Nice to have an investment in property
More independent - only yourself to rely on
Inheritance for family
More privacy to do what you want when you want within reason
It's a lot cheaper - mortgages are cheaper than rent
Not just throwing money away
You have an asset at the end of the day
You are in control and fully independent
Build up your estate for the future
Feeling that it's yours even if it's the mortgage company's
Won't be turfed out by the landlord
Step on the social ladder
More control over living space
Still a reasonable investment
Gives you a lot more choice

It is perhaps significant that a number of respondents, when asked to consider the advantages of renting had not answered the question at all or had answered 'none'. Amongst those who had responded, as might be expected some of the perceived disadvantages with buying your own home were translated into advantages of renting your home (see Box 3). Thus, renting was considered not to be such a long term obligation, involved less responsibility, particularly with upkeep and maintenance, and required less commitment. Flexibility and freedom to move also emerged as an advantage. One respondent commented that there was no need to worry about selling. The issue of selling was mentioned as a disadvantage of buying your home - 'if you want to sell you get can get stuck', or 'you can get caught in a chain or not be able to find a buyer'.

Box 2

What are the disadvantages ?
Illustrative verbatim statements

If you want to move trying to sell can be difficult
General costs over and above the mortgage
Maintaining the property
Always a risk of losing money
Loss of freedom to move to another area
If you borrow a lot of money it's a big commitment. If you lose your job or prices go down it can be a great burden
Subject to the fluctuating property market
Chance that prices can go down
Can't think of any
Worry about mortgage
Can't move when you want
Its a very big commitment - if you want to move you can get stuck in a chain and you can wait forever
Not having a secure job
Having a family and not enough money
If you want to sell you might not be able to find a buyer
More tied and committed than you would be renting
Financial difficulties
Costs seem never-ending
Trying to sell when you want to move
There are none
Losing your investment
Tied to one house

The disadvantages of renting tended to be financial and the notion of paying rent as 'dead money' or a 'waste of money' was frequently expressed (Box 4). Renting was also thought to be restrictive, with the landlord 'in charge of your life', poor quality and expensive.

Box 3

What do you consider to be the advantages of renting your home? Illustrative verbatim statements

Don't have to worry about maintenance
Move more quickly
Move when you want
Repairs paid for by landlord and not yourself
No responsibility
Not tied to a mortgage
No worry about selling it
Can move without hassle of legal fees, etc.
No worries, you just call the owner
You are not tied down
Only got the rent to think of
Not much
Useful if saving up to build up a deposit to buy or if sold and looking for another house
Probably a cheaper option in London
Not such a long term obligation
No worry of losing investment and negative equity

Images of owning and renting

As a way of delving further into contemporary images of housing tenure, respondents were pressed to provide up to five words or phrases they associated with home ownership and renting (Box 5). Home ownership was most likely to be associated with positive words or phrases like *independence, comfort, sense of security* but there was also a range of negative responses. such as *expensive, responsibility* and *worry* . Other words included *aggravation* and *constant repairs*. These responses did suggest a degree of disenchantment with home ownership but it was still phrases such as *pride in ownership, social status* and the idea of *'building up your estate'* and *passing the property on to your family* which were most common.

Box 4

And what are the disadvantages?
Illustrative verbatim statements

Paying money to someone else
Paying someone else's mortgage
Having to live from contract date to contract date
Less security
Some property is a bit yukky
Can't have property as you would like it
It's never your own
Problem if you have a bad landlord getting repairs done
High rent charges
It's dead money
Outlay with no return
Waste of money
Difficulties with landlord
Less likely to be able to get credit
Landlord can be in charge of your life
Tend to pay over the odds in rent
Restrictions - can't stick things on walls etc.
Nothing to show for your money
Most don't allow dogs

Renting was associated with more negative words or phrases like *insecurity, lack of privacy, expensive, dead money, paying someone else's mortgage, never feels like your own home*, but was also seen more positively in terms of *flexibility, no commitment, landlord does repairs* (Box 6). One respondent said simply, '*don't do it!*'

A more summarised and ranked version of the above responses from all questionnaires is shown in tables 7.8 and 7.9. As can be seen, home ownership continued to be strongly associated with stability, security, freedom and general well-being. This was balanced against anxieties about mortgage and other costs and more general, unspecified concerns about the

Box 5

Words and phrases associated with home ownership
Investment
Planning for the future
Pride in ownership
Security for family
Your own
Comfort
Social - get to know neighbours
Step on social ladder
Happy
Having property to sell
Having property to pass on
Paying to own
Choice of where to live
Freedom
Permanence
Rewarding
Commitment
Renovation
Mortgage
Decorating/Repairs
House prices
Estate agents
Building societies
Cash outlay
Takes all your money
Dreading the postman
Poverty
Takes a lot of your time and energy
Worry
Loss of investment
Worry of meeting monthly payments
Weight on your shoulders
Too easy to get mortgages
Always a headache
Bills
Expensive
Insurance
Debt and deposit

responsibilities involved. For the majority of respondents, for all the potential and in some cases recently experienced difficulties of house purchase and ownership, it was clearly a price worth paying. In passing, it

---

Box 6

Words or phrases associated with renting

Dead money
Deposit bond
Eviction
Rent
Landlord
Expensive
Insecure
Never your own
Exorbitant rent
Watercress in the carpets
Mould
Damp
Lack of privacy
Less enjoyment
Lack of incentive
Do as you are told
Bad neighbours
Poor area
Being controlled
Dependence
Financial vulnerability
Never stabilising
Councils
Small Ads
Temporary
Second hand houses
Feeling of transition
Not as much worry
Easy to move
Freedom
Flexibility
No commitment

---

is perhaps worth noting that very few respondents mentioned words or phrases associated with social status in relation to home ownership.

Table 7.8
Proportion of all mentions of words and phrases associated with home ownership

	%	Rank
Stability, security, permanency well-being	28	1
Mortgages, bills, debts, repair costs	23	2
Freedom, independence, choice	18	3
Worries, anxieties, commitments, responsibilities	18	3
Investment, assets, wealth	13	5

Table 7.9
Proportion of all mentions of words and phrases associated with renting

	%	Rank
Flexibility, freedom, no responsibilities	21	1
Wasted money, lack of investment, lack of asset	19	2
Bad housing conditions, bad environment, noise etc.	17	3
Landlords, dealing with landlords, difficult landlords	12	4
Insecurity	12	4
Lack of freedom, restrictions	11	6
Expense, high rents	8	7

The image of renting in whatever form was generally very negative. It also offered freedom in the sense of flexibility, ease of movement and a lack of responsibility for repairs and upkeep of dwellings. But these traditional qualities of renting (particularly private renting) were balanced against a pervasive image of low quality housing, difficult landlords and rents representing dead money. It is this negative image of renting, and particularly private renting (given the housing experiences of most of our respondents which did not include social housing) which helps to explain the resilience of home ownership despite the problems experienced by many of our home owners.

Concluding comments

It is evident that even after a prolonged recession, and among those home owners most seriously affected, general attitudes towards home ownership remain positive. In 1997 most respondents thought house prices would recover although expectations about dwellings as investments appeared more muted. The traditional optimism associated with home ownership was still evident amongst many of the respondents. There was still the belief that investment in 'bricks and mortar' offered security and something to pass on to the next generation. However, this optimism was tempered by a recognition that there were no guarantees about rising housing prices and stable mortgage interest rates. Home ownership could be a constraint, limiting mobility and reducing flexibility and freedom, as well as a burden, a responsibility and a major commitment. It seemed that a new set of concerns had emerged about home ownership which were about loss of investment, negative equity, and an inability to sell the property and move easily. There was also a recognition that deposits and transaction costs could no longer be taken care of by house price inflation and were thus significant constraints on mobility for some or at least a factor which would have to be weighed against, for example, commuting costs. This may have been one of the factors which pushed the process of property purchase and sale up the political agenda and led the 1997 Labour government to instigate a review of transaction cost. It was also evident, however, that a significant proportion of households had moved out of negative equity.

There are a number of issues which emerge from this exploration of the attitudes and behaviour of these overlapping subgroups in the sample. Most obviously, the impacts of the changes in the owner occupied market coalesced with other aspects of family and employment change to produce different consequences for different groups. As has been said elsewhere, negative and limited equity has for the majority of those affected represented a new, inconvenient and worrying problem but one which only became a serious issue when combined with other financial and family pressures. It was growing families in the 30-44 age range which had been most frustrated in terms of mobility and most constrained by the financial costs of having a young and growing family. These constraints had been greatest in households with only one earner and with limited earnings which may have been a consequence of having (more) children. The response of many households in this group, however, was to have invested more in their present home as perhaps the only available way of producing more or higher quality space in the absence of sufficient resources to move. It was those who physically could not adapt or expand their dwelling (e.g. those in flats or those who could not afford to) who had experienced the worst problems.

The younger cohort of owners had different and shorter experiences of the housing market and were more likely to be pessimistic about house price trends. For this group the dominant experience of the housing market had been one of rapid boom followed by long-running bust. They saw themselves as a cohort or generation which had suffered financial loss through home ownership, whereas older ones, had made significant gains. Those following the negative equity cohort were seen as being in a better position. Although they were seen as unlikely to benefit from the price gains of the past, they had been able to secure more or better housing for their money. Older home owners were more likely to expect a return to what they had come to regard as more normal conditions and to see recent trends as merely an extreme version of previous ups and downs in property prices. It remains to be seen whether a sustained recovery in the housing market will affect the attitudes of younger, first time buyers who entered the market in the late 1980s.

As identified through our earlier in depth work there was also a group of owners which had been in a position to save to reduce negative equity or to build up sufficient funds to pay for transaction and other costs of moving house. This group appeared to be most typically established households with more disposable income and to be single or childless couples. For some this had involved a degree of hardship. For others, delayed mobility in the housing market had produced surplus funds which under different market circumstances would have been invested in larger, better quality accommodation.

The overall picture to emerge was of a high degree of frustration with the way in which previously held expectations about housing adjustments and mobility had been compromised by the depth and longevity of the recession. This general frustration was, however, only associated with more deep

121

seated problems and perhaps a longer standing shifts in attitudes and behaviour for a minority of lower income, financially vulnerable home owners.

8 The rise and fall of negative equity in Britain

This book has focused on a particular period in the history of the British housing market. Drawing on three separate but related pieces of original research, and on the work of others as appropriate, it has highlighted the impact of the post 1989 recession on homeowners caught by an unprecedented fall in nominal house prices. The main emphasis of the book has been therefore on individual experiences and coping strategies in a situation of unpredictability and uncertainty, and the impact of these experiences on broader attitudes to home ownership.

The cohort of households which purchased dwellings in the late 1980s and early 1990s experienced a very different kind of housing market to the one to which most existing owners trading up and new purchasers had become accustomed. Cyclical rises and falls in real house prices had been part of the post war experience of home ownership but housing had generally been a safe and sensible investment because house price inflation had tended to outstrip general price inflation. When house prices had fallen in real terms this had been concealed by general price inflation and continuing rises in nominal house prices. The sharp fall in nominal house prices of the late 1980s followed by a long and apparently deeply entrenched recession in the property market during the 1990s were novel experiences for house buyers and existing owners.

Longstanding views and expectations amongst home owners generally about the process of housing adjustment and residential mobility, and about the prospect of continuing equity gain, had to be revised in the light of events in the property market. However, there were more deep seated problems and perhaps a longer standing shift in attitudes and behaviour for a minority of lower income, financially vulnerable home owners. There was a strong impression of polarisation among households with negative equity.

123

At one extreme was the two earner, professional household which may have experienced substantial negative equity over an extended period, but in the context of rising real household income. For this type of household the problems of the housing market were generally inconvenient rather than presenting major difficulties. Some longer distance job mobility had been affected but in most cases a shorter distance move was merely delayed while waiting for house prices to recover. In some cases, mobility was facilitated by entry into the private rented market as both landlord and tenant. With the recovery in prices most of these households will now have moved out of negative equity. Some were even able to buy themselves out of it by repaying some of their mortgage from income or savings.

At the other extreme, however, were lower income households with more limited and more vulnerable earnings, buying lower value or small properties and where negative equity often coalesced with other financial or domestic difficulties such as redundancy or relationship breakdown. It is this group, a small minority of those caught in the late 1980s recession, which is most likely to remain as longer term casualties in the aftermath of negative equity. Between the two a much larger group were knocked a few rungs down the ladder but did not then fall off. Bearing in mind that the impacts of negative equity varied substantially we can summarise the characteristics of the typical households affected as follows:

- households which purchased a dwelling in the 1987-90 period during the peak of the boom and in its aftermath before it became clear that prices would continue to fall over an extended period;
- single persons in full time work or two earner households, both working full time;
- alternatively in the early stages of the family life cycle; with young children;
- mainly but not exclusively first time buyers;
- as likely to be in professional or managerial employment as in manual categories;
- living in the south of England;
- spread across the housing stock but concentrated in areas with high levels of completions in the 1987-90 period.

As regards general housing market dynamics, the most obvious evidence of the changing fortunes of home ownership was the substantial fall in the number of transactions. This was reflected in both our surveys and in the qualitative work. Many households with negative equity had either tried to move and failed to, or would have liked to have moved but either did not

124

think it was worth the effort or could not afford to do so. Moreover, even among those households which were not in negative equity at the time of interview, almost a fifth felt that moving would present major financial difficulties. Clearly, many households felt the need to build up positive equity rather than simply have their negative equity removed by rising prices before their mobility problems were eased. Awareness of transaction costs grew as these were no longer subsumed by rising prices. Specific financial packages introduced by lenders to ease mobility among households with negative equity appeared to have had a very marginal impact.

In some cases, restricted mobility had clearly involved more severe difficulties. Households had delayed starting or extending families, presumably because of space restrictions in their current dwelling, financial problems or combinations of the two. The impact on housing investment strategies was ambiguous. Some households had chosen to invest more in their current dwelling because of delayed mobility. Others took the view that the changed housing market conditions had influenced them to spend less on their home.

Amongst households more generally there was an apparently more cautious approach to house purchase with a greater emphasis on housing as a consumption rather than as an investment good. This new caution was reflected in an apparently greater reluctance by some households to invest surplus or windfall gains in housing and less inclination to trade up-or perhaps to trade up to a lesser extent. There was also a greater awareness of the costs and responsibilities associated with home ownership, even in the service sector and the South East. Uncertainties about income and job security also came to prominence for the first time in the 1990s recession and these further undermined confidence and willingness to buy or to trade up.

Business as usual?

Over the period of the recession in the British residential property market, research by ourselves and others has provided substantial evidence about the geography of negative equity in Britain, the characteristics of those most affected and the broader effects on the housing market. The obvious question is whether it matters now other than as a description of the residential property market in Britain at a particular historical conjuncture. Hasn't negative equity been and gone? Are we not back to business as usual? The region of the country, the south east, which was most affected by house price falls has also experienced the strongest recovery. A house price report from the Nationwide Building Society (1997) observed that "Nearly all regions saw price increases, but the largest gains continued to be recorded in the south east. Over the year, price inflation remains highest in London (17.5%) ... In general, price rises are weaker as one looks north and only London appears to have seen a significant rise in house sales over recent months" (p.1). The number of households with negative equity has declined dramatically and as a phenomenon or policy problem it looks set to disappear in the near future. But is it?

There are a number of dimensions to this issue which bear consideration. First, could it happen again? One thing most analysts agree on is uncertainty and the need for greater caution among borrowers and lenders. Bootle (1996), for example, in his analysis of the future of housing markets in an era of low or zero inflation suggests that substantial nominal house price falls could well happen again creating another phase of widespread negative equity. More generally if there is an extended period of low inflation it would be expected that house prices would be affected in the same way. Market commentaries are already indicating a much more complex picture of price inflation and recovery than has been the experience of the past-and the economic crisis in east and south east Asia casts a major new shadow over western economies. If there is a current recovery of house prices in Britain it is certainly not a simple bust to boom. It is uneven, precarious and involves a relatively low level of transactions. On the other hand there are no long-term grounds for assuming that another cycle of house price boom and bust does not lie ahead, even if not in the short-term. Rising incomes and increasing demands for housing arising from demographic pressures including the inexorable trend towards household fragmentation and flexible living arrangements, set in an increasingly restrictive planning climate restricting new construction, may well lead to the same pressures which gave rise to the late 80s boom.

Second, and related to the above, are certain areas, dwellings and households likely to be left behind? It does seem that the broader analysis of general societal polarisation and division is finding particular expression in the property market. Differentiation and fragmentation in the home ownership sector is increasingly evident (Forrest, Murie and Williams, 1990). This is well illustrated by a number of accounts of the state of the residential property market which have recently appeared in Sunday newspapers. In one weekend (13/14 June, 1998) *The Guardian* (13.6.98) headlined a feature on property prices, "Is your house hot or cold?", referring to extremes of high or low/negative house price changes. Examples were given (drawing on a report from the Land Registry, http://www.open.gov.uk/landreg/home.htm) of Bristol where "Georgian and

Regency houses around Clifton were almost doubling in value in a year, and on the opposite side of the city small redbrick terraces in former industrial areas were declining in value by 65 per cent" (p.14-15). The next day *The Independent* had a front page headline "Property bubble bursts" (1.6.98) and *The Observer* featured a piece (again on the front page) "Basement flat sells for record £4.5 million" - a reference to a cash sale in a new prestige block of flats in London's Hampstead. Not only does this illustrate the continuing centrality of house prices in popular discourse but also the pervasive confusion of what is actually happening in the market. In some areas there is rampant house price inflation, a lack of properties to buy, affordability problems, gazumping and falling house prices. In others dwellings change hands for a few thousand pounds at auction or are simply abandoned. 'Location, Location, Location' is the prevailing wisdom. While the Nationwide Building Society (Nationwide, 1997) commented that negative equity had 'come down sharply' and estimated that the number of households affected had halved between December 1996 and the end of June 1997 - from 810,000 to 410,000 it qualified its generally sanguine prognosis by emphasising that "deeply entrenched pockets of negative equity could remain for some time, typically among owners of certain types of property at the lower end of the market which may be hard to sell in anything but 'boom conditions' (p.4). For some therefore, it may be a very long wait to be lifted out of negative equity and this may represent a diminished but deeply entrenched problem with, at the extremes, prospects of residential abandonment and blight. If we see negative equity anew in the future, it may be <u>localised</u> booms and slumps as much as national ones which bring this about.

Third, to what extent have individuals and households suffered long term damage? General quantitative estimates of the number of households in negative equity convey little about the material and psychological impacts of the property market recession on homeowners. For the majority of those in negative equity the fact that their mortgage debt exceeded the market value of their property was merely an unexpected inconvenience - worrying but not representing a crisis provided that mortgage payments could be met. An expected move to a larger or more expensive property may have been delayed but this was not a life changing event. In some cases job moves were affected because of difficulties of selling but especially for those in higher status employment and with major companies or in the in public sector, employers often stepped in with schemes to alleviate difficulties. But for others negative equity combined with factors such as unemployment, rising interest rates, relationship breakdown, reduced mobility, delayed family formation and general indebtedness to produce major difficulties. In our research we came across households with difficulties which would not simply melt away with a general price rise. Often they were living in precisely those kinds of properties (small, low value flats or houses on large, new estates; badly maintained inner city terraces; ex-council flats) which may be continuing to fall in value or being left behind in relative terms. In some cases the coping strategies they had adopted to deal with their difficulties such as borrowing from friends or relations would themselves have medium to longer term consequences for their housing careers. There were separated couples who could not sell their joint home. There were couples wishing to have (more) children but unable to do so for lack of space. For example, there was evidence to suggest that some one in 7 households with negative equity had delayed having (more) children because of the state of the property market (Forrest, Kennett and Leather, 1998). There were those who could not move to take up a job opportunity because of difficulties of selling or commuting long distances because they could not move. At the extremes we came across families of four living in one room bedsits. Those who had been on the margins of affordability in the late 1980s found themselves at the sharp end of unsaleability and falling values in the 1990s. For many the aftermath of negative equity is at the very least a set of changed expectations and plans regarding the timing of children, house moves, savings and jobs. The public and policy debate may have moved on but many instances of private misery remain.

Fourth, there is the aftermath of negative equity in terms of changed attitudes towards home ownership more generally. Do people feel differently - more cautious about trading up, disillusioned, more inclined to rent rather than buy, more likely to rent for longer before buying and so on? Have a particular cohort, the late 1980s casualties, a different set of attitudes from previous cohorts of home owners or those who will follow? We have suggested that there appears to be greater circumspection among home owners generally and more qualified expectations of home ownership. But the simple answer is that there is little evidence of any profound change of attitudes to home ownership in principle. This resilience, sometimes in the face of some thoroughly miserable experiences of the tenure, can be explained mainly because of the lack of attractive alternatives. Home owners, it seems, judge their own experiences of the tenure not so much in relation to other home owners but in comparisons with renting, and particularly private renting. It may be that this has a particularly British complexion and reflects the nature and use of private landlordism in Britain-often a tenure of transiency, associated with poor quality accommodation.

At the height of the property recession in Britain there was some speculation that the gloss had finally come off home ownership. Another episode of nominal price falls amidst rising unemployment and repossessions could well inflict further damage. And there is evidence of first time buyers staying longer in their first properties and of a marked reduction among younger households entering the tenure (Office for National Statistics, 1996). It is this latter group which have chosen to enter the private rented sector in greater numbers than previously and to remain there for longer. To that extent housing market behaviour *has* been modified by the recession in general and the threat and experience of negative equity in particular with greater caution about trading up and debt exposure and delayed entry by younger people into home ownership. Indeed, some of the delayed entry into home ownership has been accommodated by the shake-out of dwellings from the tenure. Some home owners with negative equity and other problems of saleability and mobility chose to rent out their properties. This process has generated an additional and for once a relatively high quality layer to the privately rented sector. Another aspect, therefore, of the aftermath of the negative equity is the way in which market adjustment of this kind has changed the tenurial composition of neighbourhoods (Forrest and Kennett, 1998). Some areas which were exclusively owner occupied now have a number of privately rented properties in their midst. It remains to be seen how far this addition to the rental sector is temporary or longer term. There is evidence of speculative purchase by individual investors who are likely to remain in the market at least until some significant capital gain can be made on their investment. What is unambiguous however is that private renting, in the British context at least, remains for the majority of tenants merely a stepping stone on the way to home ownership. Choko (1995) has made the point that to explore attitudes to home ownership merely in relation to investment expectations or with some rational model of market behaviour is perhaps to miss the point. The continuing attraction of home ownership, even in the wake of rampant negative equity, repossessions and constrained mobility, is to some extent explained by negative rather than positive factors. "To be an owner, or rather to become an owner ... is above all not to be a tenant anymore" (Verrett, quoted in Choko, 1995, p.147).

Unlike previous crises or major policy problems in housing, the phenomenon of negative equity affected neither the poorest households nor the poorest regions in the country. On the contrary it was a problem concentrated in the economic and home owning heartlands of Britain, the south east in and around Greater London. As the problem has diminished with some sustained market recovery there is a tendency to measure the continuing presence of negative equity in simple numerical terms. Such measures, however, underestimate the gravity of the difficulties faced by some households and the way in which the problem of unsecured housing debt (negative equity) often interacted with and exacerbated other personal and financial difficulties. A particular cohort of home owners experienced a very different set of market circumstances than had their predecessors. And within that cohort, there is a group of households and dwellings, which suffered from disproportionately high falls in property values and severe problems of saleability and mobility. It is this group, and here we are referring mainly to lower income home owners, which is in danger of being left behind as market conditions change. Not only have these ones fallen equity accumulation stakes but they have also in some cases been overtaken by their successors – snapping up bargains in the sales of the early 1990s.

General attitudes towards home ownership appear to have remained positive with the traditional associations of the tenure with stabilty and investment. But the world does appear to have changed. Divisions within home ownership have hardened with a more differentiated pattern of price movements. In a less certain market the best dwellings in the most sought after locations acquire additional premiums as positional goods. Conversely, the least attractive properties in the least attractive locations, often those on the affordability margins of the 1980s boom, lag behind as buyers exercise greater scrutiny and choice and leave less to chance. If lessons have been learnt it is that prices can fall, and can fall sharply, and that there are casualties which may not recover. The British experience also resonates with (and is linked to) past and current patterns of house price volatility in other countries. At present the economic crisis in South east Asia has undoubtedly plunged a new cohort of home owners into negative equity producing similar if not more severe policy dilemmas for governments and chronic and unfamiliar difficulties for individuals and families.Commenting on the growth of negative equity in Hong Kong and the difficulties of selling properties, the Economist (1998) observed that when people are asked what they think about it they typically 'reply along the lines of "The government should do something"' (p.51). In practice, of course, governments can do very little and just as in Britain most home owners will simply have to wait for better times to come. And just as in Britain different households will have different capacities to cope with risk and instability in the housing market. Negative equity is merely one new risk (albeit one which for a time at least grabbed the headlines), among many which households in a variety of national contexts have to cope with through market and informal means. In that sense it is appropriate to see negative equity as a significant but by no means the only or even the most threatening problem in contemporary life paths which are likely to be increasingly punctuated by multifaceted experiences of unpredictability and insecurity. How households – and the housing market – respond to this climate of insecurity and the resulting impact on the demand for home ownership or other tenures remains to be seen.

Bibliography

Bank of England (1992) Negative equity in the housing market, *Bank of England Quarterly Bulletin*, pp.266-269.

Boleat, M. (1994) The 1985-1993 Housing Market in the United Kingdom, *Housing Policy Debate* 5, pp.253-274.

Booth, P. and Crook, T. (eds.) (1986) *Low Cost Home Ownership*, Aldershot: Gower.

Bootle, R. (1996) *The Death of Inflation. Surviving and Thriving in the Zero Era*, London: Nicholas Brealey.

British Market Research Bureau (1991) *The Property Market*, BMRB for the Council of Mortgage Lenders and the BBC Money Programme.

British Market Research Bureau (1993) *Housing and Saving*, BMRB for the Building Societies Association.

Choko, M. (1995) 'Home owners: richer or not?' in Forrest, R. and Murie, A. *Housing and Family Wealth Comparing International Perspectives* London: Routledge.

Coles, A. (1992) 'Causes and characteristics of arrears and possessions' *Housing Finance*, 13, pp.10-13.

Cooper, A. and Nye, R. (1995) *Negative Equity and the Housing Market*, Social Market Foundation Memorandum, Social Market Foundation, London.

Council of Mortgage Lenders (1996) Negative Equity: Outlook and Effects, *Council of Mortgage Lenders Research* No.8, London.

CSO (1992) Social Trends 22, London: HMSO.

Department of the Environment (1993) *English House Condition Survey 1991*, London: HMSO.

Department of the Environment (1994) *Housing Attitudes Survey*, London: HMSO.

Doling, J. and Ruonavaara, H. (1996) Home Ownership Undermined? An analysis of the Finnish case in the light of British experience *Netherlands Journal of Housing and the Built Environment*, 11, 1, pp.31-45.

Doling, J. and Stafford, B. (1989) *Home Ownership: The Diversity of Experience*, Aldershot: Gower.

Dorling, D. and Cornford, J. (1995) Who has Negative Equity? How House Price Falls in Britain Have Hit Different Groups of Home Buyers, *Housing Studies*, 10, 2, pp.151-178.

Dorling, D., Gentle, C. and Cornford, J. (1992) *The crises in housing: disaster or opportunity*, CURDS Discussion Paper 96, October 1992.

Earley, F. (1997) Recent Developments in the Housing and Mortgage Markets, *Housing Finance* No.34, Council of Mortgage Lenders: London, pp.5-8.

Edel, M., Sclar, E. and Luria, D. (1984) *Shaky Palaces: Homeownership and Social Mobility in Boston's Suburbanization*, New York: Columbia University Press.

Fallis, G. (1995) *Structural Changes in Housing Markets. The American, British, Australian and Canadian Experience*, Paper presented at XXI World Congress, International Union of Housing Finance Institutions, London.

Ford, J. (1994) *Problematic Home Ownership*, Joseph Rowntree Foundation.

Ford, J. and Kempson, E. (1995) Mortgage arrears and possessions: prospectives from the borrowers, lenders and the courts, HMSO, London.

Forrest, R. and Kennett, P. (1998) 'Re-reading the city: deregulation and neighbourhood change' *Space and Polity* 2,1, pp.71-83.

Forrest, R. and Murie, A. (1990) *Selling the Welfare State*, London: Routledge.

Forrest, R. and Murie, A. (1994) Home ownership in recession, *Housing Studies* 9, pp.55-74.

Forrest, R. and Murie, A. (1995) *Housing and Family Wealth: Comparative International Perspectives*, London: Routledge.

Forrest, R., Gordon, D. and Murie, A. (1996) 'The position of former council homes in the housing market' *Urban Studies* 33, pp.125-136.

Forrest, R. Gordon, D., Pantazis, C. and Leather, P. (1995) *Home Ownership in the United Kingdom: The Potential for Growth?*, London: Council for Mortgage Lenders.

Forrest, R., Kennett, P. and Leather, P. (1994) *Home Owners with Negative Equity*, Bristol: Policy Press.

Forrest, R., Kennett, P., and Leather, P. (1997) *New problems on the Periphery? Home Owners on New Estates in the 1990s*, Policy Press: Bristol.

Forrest, R., Kennett, P. and Leather, P. (1998) *Attitudes to Home Ownership in the 1990s*, Final Report for the Department of the Environment, Transport and the Regions, Bristol: SPS.

Forrest, R., Murie, A. and Williams, P. (1990) *Home Ownership: Differentiation and Fragmentation*, London: Unwin Hyman.

Gentle, C., Dorling, D. and Cornford, J. (1994) Negative Equity and British Housing in the 1990s: Cause and Effect, *Urban Studies* 31, 2, pp.181-199.

Hamnett, C., Harmer, M. and Williams, P. (1991) *As Safe As Houses*, London: Paul Chapman.

Harvey, D. (1978) 'The urban process under capitalism', *International Journal of Urban and Regional Research* 2, pp.101-131.

HM Land Price Registry (1997) *Residential Property Price Report*, London.

Holmans, A.E. and Frosztega, M. (1996) *Negative Equity*, Department of the Environment: London.

Jowell, R. et al. (eds.) (1997) British Social Attitudes 14th Edition, Ashgate: Aldershot.

Karn, V., Doling, J. and Stafford, B. (1986) 'Growing crisis and contradiction in home ownership' in Malpass, P. (ed.) *The Housing Crisis*, London: Croom Helm.

Karn, V., Kemeny, J. and Williams, P. (1985) *Home Ownership in the Inner City: Salvation or Despair?*, Aldershot: Gower.

Kempson, E. and Ford, J. (1995) *Attitudes, Beliefs, and Confidence: Consumer Views of the Housing Market in the 1990s*, Council of Mortgage Lenders, London.

Kennedy, N. and Andersen, P. (1994) *Household Saving and Real House Prices: An International Perspective*, Working Paper No.20, Bank for International Settlements, Basle.

Kosonen, K. (1995) *Pohjoismaiden asuntomarkkinat vuosina 1980-1993. Vertaileva tutkimus*, (Nordic Housing Markets in 1980-1993. A Comparative Analysis) Helsinki: Palkansaajien tutkimuslaitos.

Lowe, S. (1992) 'The social and economic consequences of the growth of home ownership' in Birchall, J. (ed.) *Housing Policy in the 1990s*, London: Routledge.

Lunde, J. (1990) *Boligejernes formuesituation - en empirisk undersogelse* Institute of Finance, Copenhagen Business School Working Paper 90.

Maclennan, D. et al. (1994) *A Competitive UK Economy: The Challenges for Housing Policy*, York: Joseph Rowntree Foundation.

Malpass, P. (1986) *The Housing Crisis*, London: Croom Helm.

Muellbauer, J. and Cameron, G. (1997) A Regional Analysis of Mortgage Possessions: Causes, Trends and Future Prospects, *Housing Finance* 34, pp.24-34.

Munjee, N. (1996) Homeownership Trends Worldwide, *Housing Finance International*.

Munro, M. and Tu, Y. (1996), *UK House Price Dynamics: Past and Future Trends*, Discussion Paper No.1, London: Council for Mortgage Lenders.

Nationwide Building Society (1997) Housing Finance Review 10 1.

Office for National Statistics (1996) *Housing in England 1994/95*, London: HMSO.

Office for National Statistics (1997a) *Housing in England 1995/6*, London: HMSO.

Office for National Statistics (1997b) *Regional Trends 32*, London: HMSO.

Pawley, M. (1978) *Home Ownership*, London: Architectural Press.

Saunders, P. (1990) *A Nation of Home Owners*, London: Unwin Hyman.

The Economist (1998) 'Down and out in the fragrant harbour' August 15-21.

Thomas, R. (1994) *Regional Housing Markets in the 1990s: The South Strikes Back*, UBS Limited, London.

Thomas, R. (1996) *Negative Equity: Outlook and Effects*, Council of Mortgage Lenders, London.

Timonen, P. (1992) *Asuntovelalliset - riskista kriisiin. Esitutkimus asuntovelkojen vuoksi maksuvaikeuksiin joutuneista kotitalouksista,* (The housing debtors - from risk to crisis. A pilot study of households in default), Helsinki: Kuluttajatutkimuskeskus.

Wilcox, S. (1998) *Housing Finance Review,* York: Joseph Rowntree Foundation.

Wrigglesworth, J. (1992) *Housing Market: more pain before gain*, UBS Philips and Drew, London.

For Product Safety Concerns and Information please contact our EU
representative GPSR@taylorandfrancis.com Taylor & Francis Verlag GmbH,
Kaufingerstraße 24, 80331 München, Germany

Printed and bound by CPI Group (UK) Ltd, Croydon, CR0 4YY

08/05/2025

01864377-0003